Linear Circuit Analysis and Drawing

Linear Circuit Analysis and Drawing

IAN SINCLAIR

NEWNES

Newnes
An imprint of Butterworth-Heinemann Ltd
Linacre House, Jordan Hill, Oxford OX2 8DP

 A member of the Reed Elsevier group

OXFORD LONDON BOSTON
MUNICH NEW DELHI SINGAPORE SYDNEY
TOKYO TORONTO WELLINGTON

First published 1993

© Ian Sinclair 1993

All rights reserved. No part of this publication may be reproduced in any material form (including photocopying or storing in any medium by electronic means and whether or not transiently or incidentally to some other use of this publication) without the written permission of the copyright holder except in accordance with the provisions of the Copyright, Designs and Patents Act 1988 or under the terms of a licence issued by the Copyright Licensing Agency Ltd, 90 Tottenham Court Road, London, England W1P 9HE. Applications for the copyright holder's written permission to reproduce any part of this publication should be addressed to the publishers

British Library Cataloguing in Publication Data
A catalogue record for this book is available from the British Library

ISBN 0 7506 1662 8

Produced by Butford Technical Publishing
Butford Farm, Bodenham, Hereford
Printed in Great Britain by The Bath Press, Avon

Contents

Preface	7
1 First principles	**9**
Linear circuit analysis	9
The computer	13
The Aciran disk	14
Directories	16
Backing up and copying files	20
Graphics cards and printers	26
Circuits and nodes	27
Components and tolerances	30
2 Aciran in action	**33**
A passive circuit	33
Saving and loading	44
Saving and using results	46
Log and linear plots	47
Editing the circuit	50
Dealing with tolerances	52
Frequency insensitive circuits	58
Effect of stray capacitance	63
Effects of tolerances	65
3 Other passive filter circuits	**67**
Analysing LC filters	75
A simple high-pass filter	76
Bandpass and bandstop filters	86
More complex RC filters	89
Using transformers	92
4 Using active circuits	**97**
Transistors and FETs	97
Transistors	98
Positive feedback	109
Audio active circuits	112
Wide-band amplifiers	117
Power amplifiers	120
Adding transistors to the set	120
FETs	122
A tone control example	122
5 Operational amplifiers	**127**
Simple inverting amplifier	129
Simple filter networks	134
Other circuits	137
Sallen & Key active filter circuits	140
Audio circuits	150

6 Last lap — 159
- Using resonant lines — 165
- Open and shorted lines — 167
- Imperfect components — 170
- Semiconductor simulations — 173
- Aciran registered version — 175
- The ACTRAN conversion utility — 177
- Notes on PSPICE — 179

7 Starting with AutoSketch — 185
- Requirements — 185
- Installation — 187
- Configuration of AutoSketch 3 — 188
- Using the mouse in AutoSketch — 190
- Limits — 191
- Measuring units — 194
- The drawing screen — 194
- The Attach option — 195
- Snaps — 196
- Using AutoSketch — 197

8 Drawing and changing shapes — 199
- Preparation — 199
- Drawing shapes — 199
- Undoing and breaking lines — 203
- Groups and parts — 206
- Parts — 208
- Creating a set of symbols — 211
- Zooms — 212
- Copy, Move and Rotate — 215
- Scale — 218
- Stretch — 219

9 Filling and dot creation — 221
- Other patterns — 227
- Arrays and the inductor symbol — 227
- Varieties of lines — 230
- Making a simple drawing — 233
- Saving files — 238
- Loading a file — 239
- Printing or plotting — 240

10 Text, layers and other items — 245
- Adding text — 245
- Text settings — 246
- Using other fonts — 247
- Typing the text — 247
- Using the Editor — 249
- Mu and ohm signs — 252
- Using layers — 254
- Properties — 256
- Minor matters — 257
- Measurements and dimensions — 262
- DXF and SLD files — 262
- Converting drawing files for DTP use — 264
- Macro facility — 265

Appendices — 267
- A: Aciran for Windows — 267
- B: AutoSketch for Windows — 269
- C: The Public Domain and Shareware Library — 272

Index — 275

Preface

This book is not aimed at the absolute beginner to linear circuits, but will be of interest to the student, the technician engineer, the interested amateur and anyone concerned with linear circuit development work. It has also many applications in any small development workshop in which costs must be kept down, and any user who has a need to find what a linear circuit is likely to do, and how best to draw the circuit, can make good use of this book.

The analysis of linear circuits used to be a long and tedious process, requiring many repetitive calculations and very prone to error even after considerable practice. It was certainly not something that amateurs would attempt, and even professional engineers often balked at the amount of effort involved. Large development organisations have used linear circuit analysis programs running on machines like the Vax for some time, but this scale of computing is well outside the reach of smaller firms, and certainly well beyond the amateur.

Similarly, the drafting work needed to draw circuit diagrams to acceptable standards used to be a considerable chore for the smaller operator, requiring time and skills which many small businesses could not spare. In this aspect of linear electronics also, the computer has come to our aid with low-cost computer drafting (CAD) packages, all of which can be used on modern desktop or portable machines.

The low-cost IBM PC clones, which now start at the price level of about £199 for a monochrome AT computer with a 1.44 Mb disk drive and 1 Mb of memory, put more computing ability on to a desktop than was available in a roomful of equipment some 15 years ago, and now a low-cost linear circuit analysis program is available. This is called Aciran, is of UK origin (so that the author is named and accessible), and is supplied as a shareware disk. This means that the disk is available for a small copying fee (see Appendix C for details) which allows you to try the program for yourself and decide if it will be useful to you. If you decide to make use of the program you can register with the author at a cost of (currently) £50, entitling you to a manual and any updates to the program, which at the time of writing has reached Version 3.1.

Versions 3.0 (and onwards) of Aciran are considerably improved as compared to the older versions, which are still available, and have no restrictions on the scale of circuit. They are capable of providing output in table form or in graph form, with the user specifying the start and end frequencies and the number of frequency steps.

I have assumed only marginal knowledge of computing, but rather more of electronics. Some assumptions have to be made – that the computer user has some experience of using the machine to run other programs using either

MS-DOS or Windows, and that the use of the important AUTOEXEC.BAT and CONFIG.SYS files are understood. There is no shortage of books dealing with the computer operating system, and if you need help in this respect the Butterworth-Heinemann list contains several titles that will be useful. In this book the setting up of the computer, the installation of Aciran and its suitability to various screen types and printers will be explained in more detail than would be required for a book oriented to computer users; the points of turnover frequency, phase shift etc. will be assumed as known in more detail.

The examples show how the circuits are set out into Aciran by naming nodes and listing the components connected between these nodes, and show the very useful graphical output. Wherever possible, illustrations are based on reproductions of the screen display of the computer (screenshots) rather than on re-drawn images. This makes the appearance less satisfactory than it would be on a paper printout, but the screen display is more revealing and is what the user will see.

From Chapter 7 onwards, the use of the AutoSketch package is described in detail as it applies to linear circuit drawing. AutoSketch is from the same stable as the industry-standard AutoCAD program, and its files are compatible with those of AutoCAD. Illustrations in this book have been made mainly with AutoSketch 3, running with MS DOS; the later AutoSketch for Windows package is illustrated in Appendix B.

For owners of computers which run the Microsoft Windows 3.1 system, both Aciran and Autosketch are available as packages to run under Windows. This has not been strongly emphasised in this book, because a great many machines are still in use which cannot, because of inadequate memory or processor, run Windows but which can run the MS-DOS versions of these programs. Since the operating principles and outputs from the packages are the same whether the MS-DOS or Windows versions are being used, the main part of this book describes the MS-DOS packages, but the Windows versions are explained in Appendices A and B. The differences are mainly in presentation; AutoSketch for Windows does contain some enhancements, but runs slower than its MS-DOS counterpart.

I am most grateful to AutoDesk for the use of AutoSketch 3.0 and AutoSketch for Windows. I am also most grateful to Rod Smith of PDSL for the introduction to the Aciran package and to James Herron, writer of Aciran, for updates to the program and discussions on Aciran. Finally, I am greatly indebted to Duncan Enright of Butterworth-Heinemann for his interest in this book, and to Henry Kelly and Petroc Trelawny of Classic FM for my background music.

Ian Sinclair
July 1993

1 First principles

Linear circuit analysis

The analysis of linear circuits is based on the well established principles of the effect of resistive and reactive components on the amplitude and phase of a sine wave for each of a range of frequencies. For very simple circuits, this can be done either by drawing phasor diagrams to scale or by the use of algebra to express the circuit reactance as $R + jX$, where X is the reactive components and j is the square root of minus 1. It is not the purpose of this book to act as a primer or revision text for these topics, but an outline of the fundamentals is necessary to show what is being achieved by the computer programs.

- ☐ Note that circuits which contain diodes, such as demodulators, are not linear circuits and most of the computer methods described here will not be able to deal with such circuits. In addition, DC operating conditions are ignored by many of the linear analysis circuits, other than to establish transistor gain. Oscillator circuits can be analysed only if the circuit is modified to break the positive feedback loop. Many linear analysis circuits will deal with components such as transmission lines in addition to resistive, inductive and capacitive passive components.

Figure 1.1 shows a generalised passive circuit that contains resistance and capacitance only. This has been taken as an example because inductors are rarely used in circuits nowadays other than in some RF discrete stages. The circuit is the most basic low-pass type of filter, and its analysis is the simplest of all. Conventionally, resistance is plotted on the horizontal axis and reactance vertically (upwards for inductive and downwards for capacitive), using scales that allow the values to be plotted with reasonable accuracy. Using the phase diagram shows that the impedance can be measured along the line marked OX (Figure 1.2), and if you are working in this way, you will have to draw the diagram for each value of frequency that you are interested in. This is comparatively simple but very tedious.

The algebraic method is no more difficult because this is a standard circuit. Looking up any reference work on linear analysis will provide

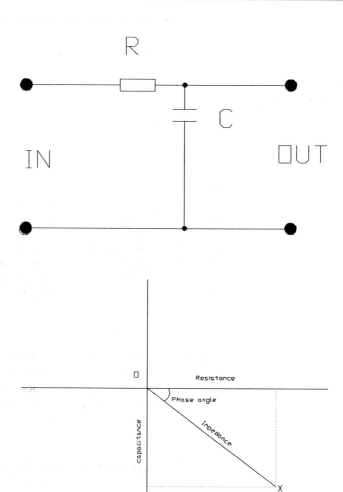

Figure 1.1 *A basic low-pass RC filter circuit to illustrate the conventional method of analysis.*

Figure 1.2 *The phasor diagram for the simple circuit, showing the quantities that represent values of impedance and phase angle.*

formulae for voltage gain (output/input) and phase angle (Figure 1.3) in which the constant ω (Greek small omega) is $2\pi f$ (frequency f). Once again, you have to work out values for gain and phase angle for each frequency in the range that interests you. A programmable calculator is a considerable help here, but the work is still tedious when you want to find the response over a range of frequencies.

When the circuit is one that is only slightly more complicated than the most basic filter in this example, the amount of work is enormously increased. For many standard circuits, the formulae can be obtained from reference books (such as the splendid *ITT Reference Data for Radio Engineers*, now published by Sams), but the amount of manipulation required becomes much greater, and in many cases you still have a lot

Figure 1.3 *The formulae for gain (amplitude) and phase angle for a standard circuit; these formulae can be looked up in reference books.*

$$\frac{E_0}{E_i} = \frac{1}{\sqrt{1+\omega^2 T^2}} \approx \frac{1}{\omega T}$$

$$\phi = -\tan^{-1}(\omega T)$$

where $\omega = 2\pi f$
$T = RC$

of work to do after working out the results of each formula. The repetitive nature of the work, unless a programmable calculator is used, means that it is very easy to make mistakes.

When the circuit is not one that can be looked up in a reference book, the analysis becomes very much more difficult. It amounts then to combining components, resistive or reactive, in series or in parallel, working out the first two, then combining with the next, and so on until the whole circuit has been covered. The aim is to express the effect of the circuit as:

$$R + jX$$

so that the impedance is the square root of $R^2 + X^2$ and the phase angle is the angle whose tangent is X/R. This analysis is long, tedious, and liable to errors. It requires a good grasp of working with complex numbers (numbers including j) and for a large circuit the effort is very considerable. The working for a simple parallel resistor and capacitor (Figure 1.4) should convince you, if you do not already know, that there is quite a lot of work involved.

Another dimension is added when a circuit contains active components. The gain of an active component converts a passive circuit, whose gain is always less than unity, into an active circuit, which can have a gain of more than unity over a considerable frequency range (the bandwidth). The gain bandwidth product of the active device needs to

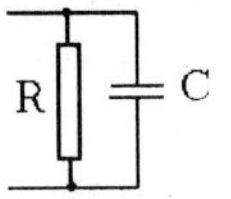

Figure 1.4 *The steps in working with a simple parallel circuit to obtain the expression in the form $A + jB$. Once in this form, the amplitude is the square root of $A^2 + B^2$ and phase angle is the angle whose tangent is B/A.*

$$\frac{1}{Z} = \frac{1}{R} + \frac{1}{1/j\omega C}$$

$$= \frac{j\omega CR + 1}{R}$$

$$Z = \frac{R}{1+j\omega CR} = \frac{R(1-j\omega CR)}{(1+j\omega CR)(1-j\omega CR)}$$

$$= \frac{R - j\omega R^2 C}{1 + \omega^2 C^2 R^2} = \frac{R}{1 + \omega^2 C^2 R^2} - j\frac{\omega R^2 C}{1 + \omega^2 C^2 R^2}$$

This is of the $A + jB$ form, so that the amplitude and phase can be found.

be considered, however, as does the effect of the impedances of the active device.

None of this need be a problem for anyone with access to a PC-compatible computer (see later), because there are now several programs which ensure that even for very complex circuits, the effort of calculating response can be performed by the computer. This makes linear circuit analysis, once possible only on very large and expensive computers, come within the reach of any user, amateur or professional. There are also circuits which can deal with bias and with non-linear circuits.

The three main programs which are currently available as low cost computer software (shareware) at the time of writing are Aciran, ACNET and PC-ECAP. ACNET costs considerably less to register but the current version is by no means so convenient to use as Aciran, since it requires you to type in each frequency (not frequency range) for which you want an analysis. In addition, the graph output makes no use of the graphics capabilities of the computer and does not provide the convenience of a single-glance display – it is necessary to shift the screen view around to follow the trace line.

The other program, PC-ECAP, is good but has not the convenience of Aciran. Since it is of US origin, the payment to the author has to be in dollars or by a cheque drawn on a US bank, neither of which is particularly convenient to UK citizens. If you like to compare programs, though, it is instructive to compare the performance of PC-ECAP with Aciran on identical circuits.

Both Aciran and PC-ECAP will provide output to a printer, though the choice of printer is rather limited in the early versions. Neither originally supported the very popular Hewlett-Packard LaserJet or DeskJet models, only the Epson and IBM dot-matrix printers. The current versions of Aciran, however, support the H-P LaserJet printer as well as the popular Epson range of dot-matrix printers, for graphical outputs.

☐ Printouts of tables of values can be obtained from any printer; it is only graph printing that requires the printer to be matched to the computer.

As well as these programs, obtainable as low-cost 'shareware' (see later), there are several programs that are commercially available, such as the excellent programs originated and marketed by Number One Systems Ltd. If your application for circuit analysis is for professional purposes you might like to look at these programs in addition to Aciran, but many users may prefer to gain some experience with Aciran before proceeding to spend more either by registering for Aciran or by purchasing a program from another source.

In this book, the displays of graphs and tables have been obtained directly from the screen of a computer running Aciran and using a screen-capture program. This makes the resolution rather coarser than a printed display, but the display is better suited for reproduction in a book. When you are using the program for yourself you will be able to obtain clearer printouts.

In addition to its obvious applications to tables and graphs of response, the use of a linear analysis program makes it possible to allow for the effect of input and output impedances and of component tolerances, something that is particularly time-consuming if done in the traditional ways. This is, however, the type of information that is particularly needed for small-scale production circuit design, so that the use of methods based on the computer is a valuable aid to anyone involved in such work. In addition, transmission lines can be dealt with; a considerable help to anyone involved in UHF design. As well as the built-in files for dealing with transistors, FETs and OPamps, there is a current generator component which can be used to simulate the action of active components much more closely at high frequencies. The main problem in using any linear analysis program is obtaining the necessary information on the active devices that are to be used, and this point will be taken up again later. Later versions of Aciran include a voltage generator component.

The computer

Any computer described as being PC-compatible – XT or AT – can be used for this work. This includes the Amstrad PC machines but not the earlier PCW machines, and it excludes machines such as the Acorn Archimedes, Atari ST, Apple Macintosh and Commodore Amiga – the use of the letter A is coincidental but a useful way of remembering that these are the incompatible machines. Note, however, that some of these machines can be fitted with a board called a *PC emulator* which allows them to be used with the huge range of software that is available for the PC machines. There are several hundreds of manufacturers turning out PC-compatible machines, and at a very wide range of prices.

In general, the older and obsolete XT design, using the 8088 or 8086 chip, should by now be very cheap, well below £200, and the more advanced AT 80286 design can now be obtained at very favourable prices. The lowest at the time of writing was £199 + VAT for a 12 MHz AT 286 machine with a floppy disk and no monitor. Much higher prices are often asked for the same machines with different badges on the front. Any machine described as a 286, 386 or 486 machine will be very well-suited to this type of work, and if it contains the chip called a maths co-processor the time needed for an analysis will be much shorter. The presence of a maths co-processor, however, is not essential, and the cost would be justified only if you were using the programs for professional purposes.

In general, for your use of the computer you will need at least a high-density floppy disk, preferably the standard 3½" HD type, a monitor, which can be colour or monochrome, and a printer, which can be any type described as Epson-compatible or H-P LaserJet-compatible. If you intend to use the machine also for computer-aided drawing (CAD) and other work, you will be advised to add a hard disk of the IDE type – a 40 Mb disk of this type costs just over £100 at the time of writing, and prices were still dropping. One of the considerable advantages of the PC machines is that the scale of production allows important add-ons like hard disk drives to be available at very low prices as compared to other machines.

The use of the later type of PC machines based on the 386 or 486 chips, along with a large-capacity hard disk is ideal – but if you do not have such a machine it is not necessary to buy one in this class simply to run the circuit analysis software. Programs run faster on such machines, but the prices are higher than for the older types. Once again, if you need very considerable power for other purposes, such a machine may be ideal and prices are falling rapidly as the chips become more readily available. The immense advantage of the PC type of machine is that compatibility has been maintained so that programs which ran on the IBM PC of 1981 can still be run on the considerably more advanced machines of today. Users of the older types of machines can also run modern software (though not all), but often at a low speed.

The Aciran disk

The linear analysis program which has been selected to provide most of the illustrations for this book is called Aciran, and in common with other shareware programs is not obtainable in the usual way from computer shops. There are many mail order suppliers of shareware programs, but the source recommended in this book is the oldest and the most experienced, the Public Domain Software Library. Its title indicates its age, because Public Domain programs were in existence before Shareware, and became common in the USA in the early 80s. A Public Domain program is one for which all copyright has been waived, so that it can be copied freely, distributed freely, and used without charge other than copying costs.

Public Domain programs arose in the USA because universities and colleges are required by law to make public property of any software that they develop. This became a rich source of some excellent programs, and the range of PD software was extended by programmers working for enjoyment, or whose ideas had not been accepted commercially. Shareware is a later concept, in which the author of software allows users to try the program free of charge, and pay for a license for continued use if the program is accepted. This allows software to be distributed at very low cost, with the author charging a modest license fee which nevertheless is more than he/she would obtain if the program were being distributed commercially.

Shareware in the UK has not progressed so well, with authors receiving very few license agreements, and a proliferation of dealers in shareware sending out some old and, in some cases, very dubious material (such as pirated commercial disks). The principle, nevertheless is a very sound one, and the chance to try out a program before paying for it is a very valuable privilege which might be lost if authors become discouraged. One interesting point is that though most shareware originates in the USA, the Aciran program and one other linear analysis program have been written in the UK, so that upgrades and support are more readily available.

To obtain the Aciran disk, contact the PDSL (address in Appendix C) and ask for disk No. 3286, Aciran. At the time of writing, this is Version 3.4 of Aciran, but if a newer version becomes available a call to PDSL will confirm the disk catalogue number and the version. If you obtain the disk from any other reputable shareware dealer (any that is a member of the Association of Shareware Professionals) you will also get the current version, but some suppliers may still be stocking the older versions.

☐ Users of Microsoft Windows 3.1 can use the Aciran for Windows program, Version 1.4, disk No. 3511. See Appendix A for details of the installation of Aciran for Windows.

Make sure that you get the correct size and type of floppy disk for your computer. Many PC machines use the older type of 5¼" floppy disk, but the more recent models use the 3½" type. Each size of disk comes in two capacities, normal or high density, measured as the number of kilobytes (Kb) of capacity (1 Kb = 1024 stored characters or codes). The low-capacity 5¼" disk will store up to 360 Kb, and the high-density disk can store up to 1.2 Mb (1228 Kb). The normal 3½" disk can store up to 720 Kb and the high-density type can store 1.44 Mb (1474 Kb). Note that 1 Mb = 1024 Kb; the use of 1024 arises because this is an exact power of two, 2 to the power 10, and is better adapted for measuring computer storage than the 1000 that is more familiar to the electronics user. A machine which is fitted with a high-density disk drive can read either type of disk in that size, but the 5¼" high-density drives will *not* write data on to 360Kb low-density disks.

The larger and more capable machines make use of hard disks. These are built into the machine rather than being inserted, run for as long as the machine is switched on, and have much larger storage capacities and faster operating speeds. A typical size range of hard disk for an AT type of computer would be 40 Mb to 100 Mb, offering much greater storage capacity and speed of access than any floppy disk. The DOS version of Aciran can be used on a machine which is equipped with only a floppy drive, but Autosketch can not, and it is almost essential on any modern machine to use a hard disk. This text makes the assumption that a hard drive is in use.

Directories

When hard disks and the larger types of floppy disks first came into use, the amount of storage that they offered was not enormous by modern standards, but it soon became obvious that new methods of maintaining a program directory were needed. The extent of the problem becomes apparent if you look at the directory of a floppy disk that contains a large number of short programs. This gives a very cluttered display which is no help in locating the file that you want to find. The user of a hard disk will probably be familiar with what follows, but if you use only floppy disks or have just graduated to the use of a hard disk, you will need this information in order to make better use of Aciran.

In addition to the problems of the number of files, there was no easy way of grouping programs in the early days. It would be logical, for example, to have a word processor and its text files, perhaps along with some printer utility programs and file-viewing utilities, in a group that would appear together when you called for a directory display. If programs can be grouped, there is also the possibility of allowing filenames to be duplicated within different groups, with no conflict. You might, for example, have a file called EXAMPLE in your word-processing group and a file called EXAMPLE in your spreadsheet group, with no possibility of using the wrong file. This is not allowed with the older type of directory system, in which the saving of a file called EXAMPLE would automatically wipe out any other file of the same name on the same disk. For a machine that makes use of floppy disks only, this is unimportant, because you would keep your spreadsheet EXAMPLE file on a disk that contained the spreadsheet program, and the word processor EXAMPLE file on the disk that contained the word processor.

The method of organising programs and data on the hard disk is built in to later versions of the MS-DOS operating system and can, in fact, be used also on floppy disks, though there is seldom any need to do so. As it happened, the older versions of Aciran used a directory system to store the files even on a low-capacity floppy disk, and the main reason for doing this was to keep files of a similar nature together. In addition, Aciran needs to use a large number of small files to hold its 'models', the data about components, and it is easier to hold such files in separate directories than to mix them with all of the other files.

The principle of the directory 'tree' is to subdivide the directory systems so that using the DIR command does not result in page after page of file listings and that saving a file does not cause problems of conflicting program names. By subdividing the directory, files can be kept in groups, as if they were on separate disks, and you can call for a directory of one group only, greatly reducing the effort that you need to spend on finding anything useful. The scheme can be operated just

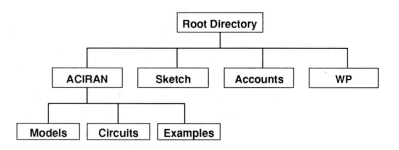

Figure 1.5 *A conventional tree diagram for a directory tree system. The directory names ACIRAN, Sketch, Accounts and WP form the first layer of subdirectories; the names Models, Circuits and Examples form a second layer.*

as easily with floppy disks, and *must* be implemented if the older version of Aciran is to run correctly even on a floppy disk.

One problem about directory trees is the name – because the diagram that everyone shows you (Figure 1.5) looks more like the roots of a tree than its branches, unless you are in the habit of looking at trees upside-down. A better way of thinking about it is as a *family tree*, which starts with one ancestor and spreads out to all of the descendants. However you like to imagine it, the diagram indicates that when you call for a directory display, all that you will get initially is the set of names that appear in the first line, called the *root names* and belonging to the *root directory*, which for a hard disk machine will be the main directory of the C:\ drive.

Some of these names can be filenames of programs or data, and they will be indicated in the usual way. Other names can be of new directories, each containing its own files. These directory names are indicated by the <DIR> following the name of the directory in place of the usual extension of a filename. In the example, these filenames are MODELS, CIRCUITS and EXAMPLES, the names that are chosen to indicate what type of files will occupy the three subdirectories. These are, incidentally, the names used for the three subdirectories of the older version of Aciran, and which can be used with advantage even with the modern version. Note that these names are not names of files but names of other directories.

The root directory

The root directory is the one to which you have immediate and automatic access when you start to use either a hard disk or a floppy. Unless you have some pressing reason to place program or data files in the root directory of a hard disk, you should always reserve it for subdirectories. In other words, when you are using the hard disk as drive C:, typing DIR should give you a list like that of Figure 1.6, in which many of the names (all of them in this example) appear between brackets, signifying a directory. On some displays, the name is shown with the word <DIR> on the same line; on other types of display the name itself appears inside square brackets. The way that the directory name is displayed depends

```
Directory of C:\

123        <DIR>    01-08-91   4:42a
ACIRAN     <DIR>    19-04-91   4:27p
ARCHIVE    <DIR>    13-10-91  11:34a
BATS       <DIR>    19-04-91   4:44p
CONVERT    <DIR>    16-05-91   9:02p
ECAP       <DIR>    11-10-91   9:30a
EDITORS    <DIR>    19-04-91   4:48p
GWBAS      <DIR>    04-10-91  10:40a
GWS        <DIR>    24-09-91   3:25p
HPSCREEN   <DIR>    08-10-91   3:08p
MSDOS      <DIR>    19-04-91   4:33p
PAGEPLUS   <DIR>    07-06-91   9:47a
PCTRACE    <DIR>    11-10-91   9:43a
PCXFILE    <DIR>    31-07-91  11:53a
SCAN       <DIR>    19-04-91   4:24p
SKETCH     <DIR>    19-04-91   4:29p
TEMP       <DIR>    19-04-91   4:37p
TIFFS      <DIR>    19-04-91   4:41p
UTILS      <DIR>    19-04-91   4:47p
```

Figure 1.6 *A directory list in which the use of brackets indicates directory names rather than filenames.*

on the operating system you are using, and the most recent versions of MS-DOS adopt slightly different methods.

You can, if you like, imagine that the root directory gives you a set of names of other disks to which you have to switch in order to obtain the program and data files that you want to use. For a floppy, the root directory will be used for the Aciran program, and the subdirectories for its data files.

The root directory is indicated in commands by using the backslash sign, \. The root directory in the C: drive is therefore indicated by using C:\, and the root directory in the floppy A: drive is indicated by A:\. The use of the backslash immediately following the drive letter (or immediately following some of the directory tree commands) always means that the root directory is to be used, but the backslash is also used as a separator to show the progression from one subdirectory to another. You have to be careful to distinguish these uses when you first start using a hard disk.

When you start work with a hard disk, you will probably start by creating a few entries in the root directory, and then probably in the first layer of subdirectories. The computer must, however, create the subdirectory names in the correct form, not as filenames, and to do this requires the use of the MKDIR command, abbreviated to MD. If, for example, you are in the directory called ACIRAN, and you want to create a subdirectory called MODELS, you do so by typing MD MODELS and pressing the ENTER key. This will cause some action in the disk drive, and when you use DIR from now on you will find the entry:

    MODELS   <DIR>

or

    [MODELS]

depending on what version of MS-DOS your computer uses. This display indicates that MODELS is a subdirectory name, not a filename. When you type DIR C:\ACIRAN\MODELS (press ENTER) you are now requesting the computer to 'follow a path' that starts with the root directory (symbolised by \) and proceeds to the MODELS subdirectory. Using this short path will allow you to find if there is anything in this directory, and will give the display of the type shown in Figure 1.7. This shows that there are no files stored in this particular subdirectory, and the only entries are the dot and double-dot.

Figure 1.7 *The DIR display for an empty MODELS subdirectory. The dot and double-dot always appear in this type of listing and the message about '2 files' really means these two directories, the current one and the one higher up the tree (the parent).*

```
Directory of C:\ACIRAN\MODELS
[.]          [..]
     2 file(s)         0 bytes
                16701440 bytes free
C>
```

The dot is an abbreviation for the current subdirectory (MODELS), also known as the 'child', and the double-dot indicates the previous directory in the path, the 'parent' directory, which in this example is the ACIRAN directory. Note that the message about '2 Files' is automatically delivered because there have been two entries printed; the MS-DOS DIR command does not distinguish between valid filenames and the <DIR> or dot displays, because a <DIR> is stored on the disk just as if it were a file. Note that these dots can be used as if they were valid directory names in commands such as CD. (CD followed by a dot).

You can create other subdirectory names of CIRCUITS and DRIVERS in the same way, starting at the ACIRAN directory and using the names that you have chosen following the MD command. This method of creating a subdirectory applies also if you are starting from another subdirectory. If you switch to the MODELS subdirectory, for example, you could create the further subdirectories of TRANSTR, FET and ICS – though you do not need to, and should not do so.

The next thing is to find how to get from one branch to another. You can change subdirectory by using the CHDIR command, abbreviated to CD, as we have noted. If, for example, you are in the root directory, then you can change to the MODELS subdirectory by typing CD \ACIRAN\MODELS. The use of CD with the slash and the name is enough to change from the root to any of the subdirectories that stem from the root. The path is the route from the current directory (the display that you get using DIR) to whichever subdirectory you want to get to. The most straightforward path is from the root to any branch that stems from that root. You do not start with the backslash symbol if you are starting from a branch along a path, only when you start from the root, or if you need to go back to the root. For example the path from MODELS to CIRCUITS requires moving by way of the ACIRAN directory, so that it needs `CD \ACIRAN\CIRCUITS`. The search for a subdirectory covers only the one layer below the directory you are presently using. It is always much easier to return to the root directory from any subdirectory, because this requires only the command `CD \`, using the backslash to mean 'root directory'.

Backing up and copying files

Disk storage for computer is reliable, but nothing is totally reliable, and the first action you need to take when you receive your copy of Aciran is to make a backup copy. This is particularly important if you work on floppy disks, because you should never use the original disks in a floppy drive machine, only copies. Backup is less important when you use a hard disk, because the working copy will be held on the hard disk, and the original disk will not be used again. Many modern computers use the 5¼" high-density disk drives, and though these drives will read a 360 Kb disk they will not write data to it. If you use your original Aciran disk and in the course of using it write data from a high-density drive, the disk will become unreadable and unusable.

Your first action, then, must be to write-protect the original disk and copy the files from the original (distribution) disk to whatever you are going to use (hard disk or floppy) for your machine.

For a 5¼" floppy disk, write-protecting the disk entails covering the rectangular cut-out at the edge of the disk with a sticky label, for a 3½" disk you need to slide the small cover open (Figure 1.8).

The methods you need to use for copying the files depend on whether your machine uses a hard disk or has only a single or double floppy drive.

- In all of the following descriptions, the word 'ENTER' is used to mean that you should press the key marked as RETURN, ENTER or with the arrow symbol ↵.

- Aciran 3.0 and following versions, unlike the earlier versions, are too large to run on a single 360 Kb drive. If you want to run Aciran 3.0 from a single floppy drive, your machine must use the high-capacity drive, which can be a 5¼" 1.2 Mb drive or either type of 3½" drive.

If your computer uses only a single floppy drive of the 5¼" or 3½" type you can make a backup copy as follows. You will need a spare disk of the correct capacity (the same as the distribution disk); formatted ready for use. This is called the *destination* disk; the original Aciran disk is the *source* disk.

BACKING UP AND COPYING FILES   21

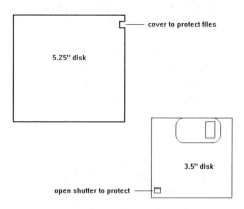

Figure 1.8 *The methods that are used for write-protecting a disk, showing examples of a 5¼" disk and a 3½" disk. Note that the 5¼" disk requires the slot to be covered for protection; the 3½" disk requires the shutter to be opened.*

☐ This assumes that the blank disk is of exactly the same size and density as the distribution disk. Do not attempt to copy a 5¼" 360 Kb disk to a 5¼" 1.2 Mb disk, nor a 3½" 720 Kb disk to a 3½" 1.4 Mb disk.

1   Place the Aciran source disk in the drive, type A: and press the ENTER key.

2   Type COPY A:*.* B: (press the ENTER key).

3   Wait until you see the message about inserting the disk for the B: drive, then remove the Aciran disk and insert the destination disk.

4   If you need to re-insert the source disk (called the A: disk) again you will be informed by a screen message, and you will get another message if the destination disk needs to be inserted.

5   You will be told when the copy action is complete. Check by placing the destination disk into the A: drive and typing the command:

   DIR

   This should display a set of filenames as in Figure 1.9.

```
A>dir /-w

Volume in drive A is 2609
Directory of A:\

-CAT2609 TXT         393 07-05-91  10:15a
ACIRAN30 ZIP      169919 28-03-91  11:29a
GO       BAT        3712 11-04-90  12:30p
PKUNZIP  EXE       23528 15-03-90   1:10a
T        COM        8191 11-12-87   3:20p
     5 file(s)      205743 bytes
                    155648 bytes free

A>
```

Figure 1.9 *The filenames on the Aciran 360 Kb distribution disk as displayed by the DIR command after the COPY action.*

☐ A single-drive machine allows its single drive to be referred to as A or B alternately.

If you are working on a machine with two floppy drives, the destination disk is placed in the B: drive, and the source disk in the A: drive. The COPY action can then take place without the need to remove or insert disks.

This action will transfer the files from the root directory of the source disk to your destination disk. If you are using a hard disk, the procedure is to create a suitable directory and then to copy the files. This is done as follows:

1. With the machine running, type C: (ENTER) and then CD \ (ENTER) to ensure that you are using the root directory of the hard disk.

2. Type MD ACIRAN (ENTER). This will make a subdirectory called ACIRAN.

3. Insert the source disk of Aciran in the floppy drive. Type A: (ENTER).

4. Type COPY *.* C:\ACIRAN (ENTER). This will copy all of the files to the ACIRAN directory.

From this point on, the methods that you must use follow along much the same lines whether you are using a hard disk or a large-capacity floppy disk.

The Aciran files are too large to be stored on a single 360 Kb floppy, and the source disk that you get from PDSL uses compression to pack the files onto a 360 Kb disk. This is done by a file-compression program which eliminates repeated characters and multiple strings and packs a large number of files into one unit. These files must be unpacked before they can be used, and you can unpack them only if you are using a disk with adequate capacity. You can ignore the following advice if you have received the unpacked files of Aciran on a 3½" disk.

☐ If your copy of Aciran comes on a 3½" disk, consists of plain (not ZIP) files, and contains a file called INSTALL or SETUP, you can set up Aciran on a hard disk simply by typing A:INSTALL or A:SETUP (whichever is used) and waiting until the program has installed the files. You can then ignore the following notes on installation.

Figure 1.9 above shows the directory of the PDSL 360 Kb disk.

The packed file is called ACIRAN30.ZIP (packed files in general will use extension letters of ZIP or ARC), and to unpack the files you need to use the program called PKUNZIP which is supplied on the same disk and which you will have copied along with the other files. To unpack the files, log into the drive or directory that contains ACIRAN30.ZIP (either by using A: or C:\ACIRAN), and check that this is the correct file by typing DIR (ENTER) when you should see the directory containing ACIRAN30.ZIP and PKUNZIP along with other filenames. The figures following 'ACIRAN' indicate the version number, so that ACIRAN30.ZIP indicates Version 3.0.

Now type the command:

PKUNZIP ACIRAN30.ZIP (ENTER)

and wait until all of the activity ceases. This unpacks all of the files – a large number of them – and reports with a screen message (using terms like Exploding... and Unshrinking...) as each file is unpacked. Figure 1.10 shows the unpacked set as they appear when you use DIR after unpacking. This consists of 570990 bytes, and you can save some space by deleting the files you no longer need, such as ACIRAN30.ZIP and PKUNZIP.EXE.

```
Directory of C:\ARCHIVE

[.]              [..]             -CAT2609.TXT    2N3819.FET      2N3821.FET
2N4338.FET       2N4393.FET       2N4416.FET      2N4417A.FET     3N163.FET
ACIRAN.EXE       ACIRAN.DOC       ACIRAN.LIB      ACIRAN30.ZIP    BC107.TRN
BC108.TRN        BC109.TRN        BC179.TRN       BC546A.TRN      BC547A.TRN
BC548A.TRN       BC556.TRN        BCY70.TRN       BF240.TRN       BF256LA.FET
EXAMPL1.CIR      EXAMPL1.NET      EXAMPL1.MAP     EXAMPL1.SCH     EXAMPL10.CIR
EXAMPL10.NET     EXAMPL10.MAP     EXAMPL10.SCH    EXAMPL11.CIR    EXAMPL11.NET
EXAMPL11.MAP     EXAMPL11.SCH     EXAMPL12.CIR    EXAMPL12.NET    EXAMPL12.MAP
EXAMPL12.SCH     EXAMPL2.CIR      EXAMPL2.NET     EXAMPL2.MAP     EXAMPL2.SCH
EXAMPL3.CIR      EXAMPL4.CIR      EXAMPL4.NET     EXAMPL4.MAP     EXAMPL4.SCH
EXAMPL5.CIR      EXAMPL5.NET      EXAMPL5.MAP     EXAMPL5.SCH     EXAMPL6.CIR
EXAMPL6.NET      EXAMPL6.MAP      EXAMPL6.SCH     EXAMPL7.CIR     EXAMPL7.NET
EXAMPL7.MAP      EXAMPL7.SCH      EXAMPL8.CIR     EXAMPL8.NET     EXAMPL8.MAP
EXAMPL8.SCH      EXAMPL9.CIR      EXAMPL9.NET     EXAMPL9.MAP     EXAMPL9.SCH
GO.BAT           J300.FET         LM124.AMP       MC1558.AMP      MPF102.FET
NE530.AMP        NE538.AMP        NE5512.AMP      NE5532.AMP      NE5534.AMP
PKUNZIP.EXE      READ.COM         READ1ST.BAT     README.BAT      README.DOC
REGISTER.DOC     RLSNOTE.DOC      STANDARD.AMP    STANDARD.TRN    T.COM
TL084.AMP        U320.FET         UA741.AMP       UA747.AMP
```

Figure 1.10 *The DIR display for the unpacked files. Some of these can be deleted as they are no longer needed. Note the large number of files after unpacking.*

If you have received Aciran unpacked, particularly if you have bought it on a 3½" HD disk, the same set of files can be copied directly from the floppy disk.

In this large set of files, only one is the Aciran program itself. There is a manual called README.DOC and a form REGISTER.DOC which you can use for registration with the author, James Herron. Many of the other files, however, are for electronic components and example circuits, and it makes the files much neater if you create subdirectories for these. This is not essential, and if you like you can place all the files into one ACIRAN directory. This is done automatically when the Windows version is installed – see Appendix A.

☐ If you are confused by MS-DOS directories, consult a book on MS-DOS, such as the *Newnes MS-DOS Pocket Book* published by Butterworth-Heinemann.

24 FIRST PRINCIPLES

For a machine with one or two high-capacity floppy disks and no hard disk, create directories on a high-density floppy disk:

1. Insert the destination disk, your working copy, into the A: drive and type A: (ENTER) to ensure that it is being used.
2. Type MD MODELS (ENTER).
3. Type MD CIRCUITS (ENTER).

This will create the correct directories on the disk. Now proceed as follows:

1. Type A:\ (ENTER).
2. Type COPY *.CIR \CIRCUITS (ENTER). This will copy all of the circuits files into the \CIRCUITS subdirectory. Check this by using the command DIR \CIRCUITS, when you should see these files. Now remove them from the root directory by typing:

   A:\ (ENTER)
   DIR (ENTER)   (to check that all the files
                       are present)

   DEL *.CIR
   DIR (ENTER)   (to check that all the CIR files
                       have gone)

3. Type COPY *.TRN \MODELS (ENTER).
4. Type COPY *.AMP \MODELS (ENTER).
5. Type COPY *.FET \MODELS (ENTER).
6. Now type DIR \MODELS (ENTER) and check that all of the TRN, AMP and FET files are present. You can then delete them from the root directory.
7. Type A:\ (ENTER) and then use DEL *.TRN, DEL *.AMP and DEL *.FET in turn, pressing the ENTER key for each command.

All of this should ensure that your disk now contains the Aciran data files in directories that Aciran can find and will use. This is important, because if Aciran cannot find its data files it will require you to specify names for these files.

For a machine using a hard disk with one or more floppy drives, the copying action must be preceded by creating a suitable set of directories on the hard disk. The methods are exactly the same as above except that C: is used in place of A:, so that the commands in order (no detail this time are:

BACKING UP AND COPYING FILES

```
C:
CD \
MD MODELS
MD CIRCUITS
C:\
COPY *.CIR \CIRCUITS
DEL *.CIR
COPY *.TRN \MODELS
COPY *.AMP \MODELS
COPY *.FET \MODELS
DEL *.TRN
DEL *.AMP
DEL *.FET
```

Now check that the contents of the ACIRAN and the other directories are as in the composite diagram of Figure 1.11. Note that you cannot normally see all of this in one screen display; several directory printouts have been combined here.

```
         C:\ACIRAN\*.*                           C:\ACIRAN\MODELS\*.*
► ACIRAN   .EXE     224,944  27-03-91     BC109    .TRN       33  19-06-88
         C:\ACIRAN\CIRCUITS\*.*                  BC179    .TRN       32  23-06-88
► EXAMPL1  .CIR         324  26-03-91     BC546A   .TRN       33  19-06-88
  EXAMPL10 .CIR       4,374  25-03-91     BC547A   .TRN       33  19-06-88
  EXAMPL11 .CIR         162  26-03-91     BC548A   .TRN       33  19-06-88
  EXAMPL12 .CIR         405  26-03-91     BC556    .TRN       34  23-06-88
  EXAMPL2  .CIR         810  22-03-91     BCY70    .TRN       32  19-06-88
  EXAMPL3  .CIR       1,296  22-03-91     BF240    .TRN       32  15-02-89
  EXAMPL4  .CIR         648  22-03-91     BF256LA  .FET       32  23-06-88
  EXAMPL5  .CIR       1,377  22-03-91     J300     .FET       29  22-06-88
  EXAMPL6  .CIR         567  26-03-91     LM124    .AMP       31  09-11-88
  EXAMPL7  .CIR         810  22-03-91     MC1558   .AMP       31  09-11-88
  EXAMPL8  .CIR         972  25-03-91     MPF102   .FET       25  23-06-88
  EXAMPL9  .CIR       1,215  25-03-91     NE530    .AMP       29  09-11-88
         C:\ACIRAN\MODELS\*.*              NE538    .AMP       29  09-11-88
► 2N3819   .FET          27  23-06-88     NE5512   .AMP       32  09-11-88
  2N3821   .FET          25  23-06-88     NE5532   .AMP       33  09-11-88
  2N4338   .FET          26  23-06-88     NE5534   .AMP       33  09-11-88
  2N4393   .FET          29  18-06-88     STANDARD .AMP       34  24-04-89
  2N4416   .FET          25  23-06-88     TL084    .AMP       32  09-11-88
  2N4417A  .FET          31  23-06-88     U320     .FET       27  23-06-88
  3N163    .FET          26  23-06-88     UA741    .AMP       28  09-11-88
  BC107    .TRN          33  18-06-88     UA747    .AMP       29  09-11-88
  BC108    .TRN          33  19-06-88
```

Figure 1.11 *The contents of all the directories are shown in this composite view (which you cannot normally see).*

The way that you use Aciran will depend considerably on whether you are using it from a floppy disk or from a hard disk. Aciran is easier to use if it is organised to keep its data in separate directories, and the processes illustrated above will ensure that these are created whether you are using a floppy or a hard disk. While you are learning to use Aciran, it is likely that the number of circuits you add to the disk will be fairly small and you are not likely to exceed the capacity of a 5¼" 1.2 Mb disk or even a 3½" 720 Kb disk. If you intend to use Aciran intensively, however, you will need to use a hard disk which will have the additional advantage of speeding up all of the disk processes such as fetching information on electronic components.

The other advantage of using a hard disk is that the files are always available. Aciran is easy to run from a floppy disk as long as all of its files are on the same disk, which must be in the drive as long as you are using Aciran. If you generate enough data to require two floppy disks you are likely to find that you need to swap disks frequently or to duplicate some data. Remember that though Aciran comes provided with a considerable amount of data for components you will need to add to this set if you want to use Aciran for serious purposes, since the set of transistors, just to take one example, is only a small selection of popular types.

Graphics cards and printers

Aciran can present its results in two ways: as a table of amplitude and phase information or as a graph or set of graphs. All PC machines can show the table on screen, but the graphs can be shown only if the computer is fitted with a graphics card, as all modern machines are. Aciran can cope with any of the normal graphics cards, but the most suitable are the Hercules and the VGA. The Hercules card is monochrome only, but the VGA type can be used either with a monochrome or a colour monitor screen. If you have a machine with a CGA card, which can display only rather low-resolution graphics, it may be possible to replace the CGA with Hercules or VGA. Such replacement is possible only if the machine is constructed in the standard way, which rules out the Amstrad PC 1512, one of the few machines that uses a CGA card. For other more compatible machines, see advertisements from Chipboards Ltd. for their VGA upgrade (graphics card and monitor combination at a very reasonable price).

Card replacement is simple on all standard machines, and is done (after unplugging all cables) by opening the lid of the machine, locating the graphics card (which has the socket for the monitor), unscrewing its retaining lug, and gently pulling the card out. The new card can then be inserted into the same slot if the card is of the same size. If you are replacing an old card on an AT-286 or 386 machine with a new Super-VGA card, you may need to look for a slot with the correct number of connectors (a 16-bit slot). Virtually all modern machines use VGA or Super-VGA display systems.

☐ Note that the Hercules card requires a digital signal (TTL) monitor and uses a 9-pin plug, but a VGA card will require an analogue monitor and uses a 15-pin plug. If a monitor is described simply as monochrome, it is usually a TTL Hercules type. VGA monitors can be mono or colour, but the resolution of a mono monitor is better unless you spend a very considerable amount on the monitor (more than the cost of the rest of the computer).

A printer is necessary if you want to keep a record of what you are working on. If you want to scrutinise the table of results closely it is much easier to do so if the table is on paper, and the same applies to graphical output. Aciran can print either tables or graphs, but graphs can be printed only if your printer is one that is supported by Aciran. Each printer needs a short program called a printer driver, and of the currently available printers the types supported by the current Version 3.0 of Aciran are those of the Epson FX type; later versions support the Hewlett-Packard LaserJet models as well. Virtually all dot-matrix printers can be set to use the Epson FX driver, so that if your printer is a dot-matrix one it can almost certainly make use of the output from Aciran. Owners of other laser printers will be able to use the H-P LaserJet emulation in the later versions of Aciran.

For graph printing, you can use either the built-in Aciran routine (press the H key when a graph is being displayed and the dot-matrix printer is ready for use) or you can make use of the facilities that are built in to the computer. Old machines with a CGA graphics card can use a program called GRAPHICS.COM to allow graphics on the screen to be printed. Hercules and VGA screens cannot be printed in this way unless you have screen-printing programs to suit (some modern versions of GRAPHICS.COM will allow VGA screen graphics to be printed). Once this has been done, pressing the PrintScreen key when a graph is displayed will result in the graph being printed. The GRAPHICS.COM program should be run by putting the name into the AUTOEXEC.BAT file (see any good book on MS-DOS), so that no separate preparation is needed. If in doubt, stick with the built-in facilities of Aciran.

If your printer is a laser type it may not be possible to print graphics in such an easy way, though later version of the PC operating system (MS-DOS 5.0 onwards) have an improved GRAPHICS program that allows the use of some laser printers. The very popular Hewlett-Packard LaserJet models can use a program called HPSCREEN which is used like GRAPHICS and will allow a graph to be printed when the SHIFT and PrintScreen keys are used together (such a combination is written as SHIFT-PrintScreen). If your printer does not fall into either of these categories you will have to find out for yourself what can be done to print out tables and graphs.

Circuits and nodes

Aciran cannot be used until it has been notified of your circuit, and since Aciran cannot read a circuit diagram (though if circuit diagrams were prepared with standardized drafting programs such as ORCAD this could be done) you need to use a system for entering the component positions and values. This system depends on identifying and numbering the circuit nodes, as you would when laying the circuit out for construction on PCB or stripboard.

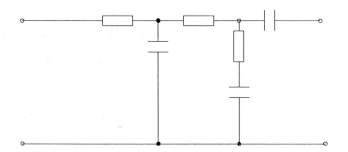

Figure 1.12 *A simple passive circuit (a filter) with input, output and a common earth line, to illustrate node location.*

Figure 1.12 shows a typical simple (in terms of components) passive circuit, with input and output and a common earth line. A node, in this sense, means a point where components join, and this will normally include the input and the output as well as the earth (ground) connection. Each node can be numbered, and the convention followed by Aciran is that the earth (Ground) node is always 0 (zero) and the input node is always 1. Other nodes can be numbered as you please.

The general rule about nodes is that no node can have more than one number, and nodes are separated by components. If there is no component between two nodes, one of the nodes is redundant. If you need to work with nodes connected, use a small resistor value such as 0.33R between the nodes. You can use whatever number you like for the output node, because this will be notified to Aciran, but you *must* use the numbers 0 and 1 for the earth and input nodes respectively. Figure 1.13 shows the same circuit with nodes numbered.

The total number of nodes in this circuit is six (numbered 0 to 5). There are also six components, but this is purely coincidental because the number of nodes does not depend in any simple way on the number of components. The point to watch in this illustration is the node numbered 4, because it is easy to overlook this one. Nodes where three or more components join are easy to spot on a well-drawn diagram because of the blob at the junctions, but this type of two-component junction is often less easy to see.

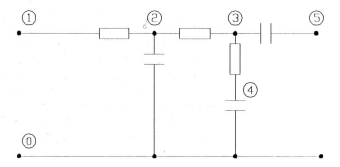

Figure 1.13 *The circuit with node numbers put into place. Only the numbers 0 for earth and 1 for input are mandatory.*

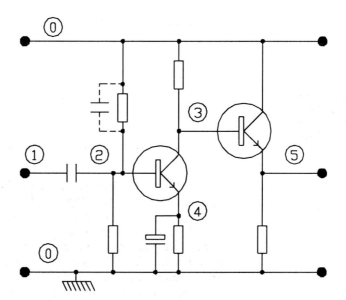

Figure 1.14 *A more elaborate active circuit with nodes numbered. The power supply line is taken as being part of the earth node, since it is at AC earth. Note the stray capacitance indicated by dotted lines.*

Figure 1.14 shows a more elaborate circuit with its nodes already numbered. This also uses six nodes numbered 0 to 5, and the point to watch here is that Node 0 is applied to both the power supply positive line and to the earth line. This is almost universal in active circuits because the power supply will normally be decoupled to earth and so is at AC earth for all frequencies that will be used. If a circuit contains any other line which is decoupled to earth (a negative supply line, for example, in an OPamp circuit) this also will be numbered as Node 0.

When Aciran analyses an active circuit, it uses a mathematical description of the active device, and this must be supplied in some form. You can opt to supply the details for yourself, but the more useful alternative is to use a ready-made device file. The form of the description is a short file which can be placed in the MODELS subdirectory, and Version 3.0 supplies some of these files (which can also be used from the ACIRAN directory if you prefer). The form of the file is simple, and you can readily add new devices for yourself – see Chapters 4 and 5 – if you can obtain the figures from manufacturers.

Figure 1.15 shows a linear IC circuit which uses eight nodes numbered 0 to 7. The positive and negative supply lines are regarded as part of Node 0 like the earth line, and you need to remember that the connection of the non-inverting input is another node, since it is connected to a component, the resistor. The node marked as 5 is also one of the type that is often forgotten.

30 FIRST PRINCIPLES

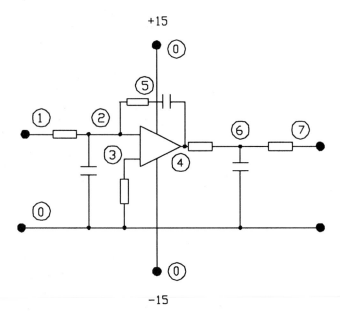

Figure 1.15 *A linear IC circuit with nodes numbered. Once again, the supply lines would be Node 0 if they had to be referred to (on a balanced circuit like this the supply lines are not connected to any other components).*

Aciran keeps data files for many of the popular linear ICs, so your circuit can be accomodated if it uses one of the set:

LM124	MC1558	NE530	NE538
NE5512	NE5532	NE5534	TL084
UA741	UA747		

There is also a STANDARD option for any OPamp whose characteristics conform to the pattern of:

Input impedance	100M
Output impedance	600R
Open loop gain	100 dB
Gain x bandwidth	100 MHz
Tolerance of open loop gain	50%

As for transistors, you can add ICs to the existing MODELS set for yourself, typing in the values from a data book if you can find one that supplies the correct parameters. The MODELS list also includes MOSFETs and the same remarks apply to these.

Components and tolerances

The components of a circuit are connected between nodes, and to describe a circuit to Aciran you need to type for each component the node connections and the component type and value. The methods that are used to notify component values are illustrated fully in Chapter 2, and in this section we shall look only at component types and tolerances. Aciran can deal with both passive and active components, as has been illustrated above, and its output tables and graphs will show the effect of tolerances and of input/output impedance.

These points are often neglected in linear analysis. Virtually every example used to illustrate analysis of a simple RC circuit in text books assumes that the circuit is connected to a source with infinite parallel impedance, and its output is to the same impedance. Adding realistic values for impedance often makes the analysis more complicated and also makes the graph of amplitude and phase quite different. Component tolerances have the same effects, and are notoriously difficult to account for if the analysis is done other than by computer – the usual manual method is to find the two 'worst-case' tolerance limits.

When each component is specified to Aciran, its tolerance can also be notified as a percentage, and this will result in Aciran producing graphs and tables that illustrate the effects of the tolerances. You can also fix a standard tolerance, such as 10%, for all passive components. For transistors, the tolerance of the h_{fe} value is the most important one for linear analysis, and is built into the mathematical model that is used – a typical value is 35% for the BC107. Tolerance of open-loop gain is the corresponding factor for a linear IC, and a value of 50% is typical. If you simply want to look at idealised results, the tolerance analysis can be turned off (see Chapter 2). Note that the tolerance value of open loop gain is seldom important for OPamps, because these devices are normally used with a large amount of negative feedback.

The input and output impedances can be specified when the circuit is analysed, and are assumed to be parallel impedances. If you do not specify anything else, these are assumed to be 100 M. Changes are made by way of a Configuration menu in Aciran (see Chapter 3).

2 Aciran in action

A passive circuit

The use of Aciran is best illustrated by examples, and the first and most obvious example to try is a simple passive circuit. The high-pass circuit of Figure 2.1 is as simple as it is possible to get, and with the component values shown and the nodes as marked, we can see what Aciran will make of it. Note that a circuit like this will behave quite differently if it is used feeding into a low output impedance, so that we shall assume in this example the use of the 100 M impedances that Aciran takes as a default unless we specify otherwise.

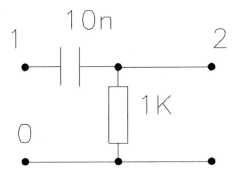

Figure 2.1 *A simple high-pass RC circuit using 10 nF and 1 K to illustrate the principles of analysis by computer.*

☐ Your circuit should be drawn up on paper, allowing space for writing, and with each component labelled in the usual way as C1, R2, Tr3 and so on. It is important to be able to identify components if you later want to change values, and Aciran allows you to enter the label names to assist in identification.

Before we start using Aciran, remember that this circuit is one that is simple enough to analyse with some quick calculations. The resistance is 1 K and the capacitance is 10 nF and using the capacitor reactance formula (Figure 2.2) shows that the response should be 3 dB down at 15900 Hz and drop at 6 dB per octave after that, assuming no loading of the circuit.

Now we can see what Aciran will make of this example.

Figure 2.2 *The capacitor reactance formula used to find at what frequency reactance is numerically equal to resistance. At this frequency of about 16 kHz the amplitude response is 3 dB down.*

$$\text{Capacitor reactance} = \frac{1}{2\pi fC}$$

and this equals R when: $\frac{1}{2\pi fC} = R$

Filling in values using units of farads and ohms:

$$\frac{1}{2\pi f \times 10 \times 10^{-9}} = 10^3 \quad \text{so that} \quad \frac{1}{2\pi f} = 10^{-5}$$

$$f = \frac{1}{2\pi \times 10^{-5}} = 15915 \text{ Hz}$$

Approx. 16 kHz for 3 dB down and 45° phase shift.

When Aciran is started, it provides the screen illustrated in Figure 2.3. This informs you of the limitations of the program to 100 nodes, 500 components and 101 frequency points in the analysis (for the version in use). This is not a severe set of restrictions for a linear-circuit program, and even if a linear circuit required more than the allowed number of components and nodes it is likely that it could be broken into sections that could be analysed separately, using the output from one section as the input for the next.

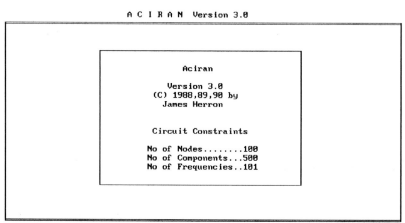

Figure 2.3 *The Aciran opening screen, indicating the limitations to 100 nodes, 500 components and 101 frequency points. Pressing any key allows the program to continue.*

As the screen indicates, you can now press any key (preferably the spacebar or ENTER key) to start the running of the program. You are then faced with the menu illustrated in Figure 2.4. This is the main menu bar. You can shift the cursor across with the cursor (arrowed) keys or by moving the mouse if your computer is fitted with a mouse. When the cursor is resting on the File portion of the menu and you press the

A PASSIVE CIRCUIT   35

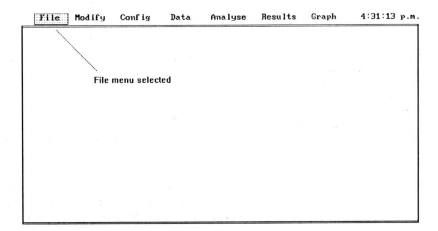

Figure 2.4 *The main menu line of Aciran with its choices, showing the File option selected.*

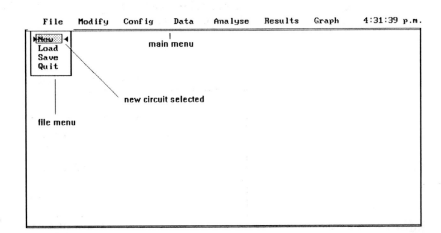

Figure 2.5 *The complete File menu, with its options of New, Load, Save and Quit. Selecting New, as shown, means that a completely new circuit is going to be described and analysed.*

ENTER key or the left-hand mouse button you will see the full *File* menu (Figure 2.5), which consists of *New*, *Load*, *Save* and *Quit*.

☐ Note that when the main menu or any of these subsidiary menus is showing you can also select by pressing the key corresponding to the letter which is in capitals, such as F for *File* or N for *New*.

The choice in this case is *File New*, since this circuit has not yet been entered. You can use *Load* to make use of a circuit that has already been prepared, such as any of the examples, but to find out how to use Aciran it is better to make use of your own circuits. Using the examples which have been provided does not provide you with the experience of marking out nodes and entering circuit components and connections. In this example we'll look in detail at how the simple high-pass circuit can be analysed.

When you select *New* you are first asked to provide a circuit description. There is no space for anything elaborate, but this does provide enough room for things like 'Simple Lo-Pass', in this case, or '6-stage LC bandpass' (later) or whatever you want to use. A good description put in here can be very useful when you are later trying to identify a circuit.

☐ This is *not* a filename, and naming the circuit at this point will not ensure that it is saved to the disk. Your circuit is saved only when you select *Save* later and provide an eight-character name for it.

You are now asked if you want tolerance entries. For a first effort like this we can do without this complication, and though the answer that is provided is Y we can press the N key to make this into a NO answer. You are then asked to select the first component of the circuit, Figure 2.6. You can choose whatever you want, but in this example I have taken the first component on the list, the resistor. This is selected by moving the shaded or coloured bar (or the arrowhead) to this line and pressing ENTER or using the mouse to move the bar and then clicking the left-hand mouse button.

You are now asked to fill in a form, Figure 2.7. This asks you for a circuit reference such as R1, C2 etc., the tolerance if tolerances are being used, and the two node numbers. In this example, R1 is the reference, the tolerance entry can be omitted because we opted not to work with tolerances, and the nodes are 2 and 1 respectively – the diagram shows

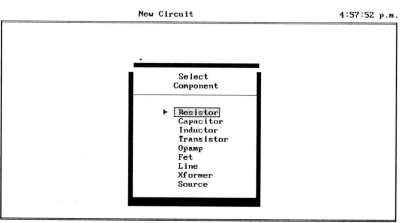

Figure 2.6 *Selecting the first component of the new circuit, in this example the resistor.*

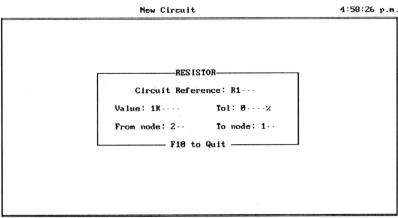

Figure 2.7 *The description for the resistor filled in, ignoring tolerances. The value can be typed as 1000, 1k, 1k0, 1K or 1K0.*

the form filled in for this example. Note that the resistance can be given as 1k or 1k0 rather than as 1000 – see below. You need to press the F10 key to escape from this form – using ENTER simply cycles you through the options again, allowing you to make any change that is needed before pressing F10.

Once this has been done, the *Select Component* form reappears and this time the *Capacitor* can be selected, Figure 2.8. When this is done, the form for a capacitor appears (Figure 2.9) and is filled in using the same methods – the value is typed as 10n, with no tolerance, and between Nodes 1 and 2.

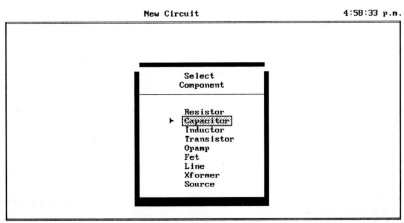

Figure 2.8 *Selecting the other component, the capacitor in this example.*

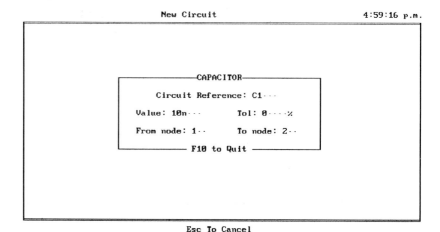

Figure 2.9 *The description for a capacitor follows the same lines as for a resistor. The value is typed as 10n; see the note on abbreviations.*

The abbreviations that Aciran will accept are:

G or g	Giga
M	Mega
K or k	Kilo
m	milli
U or u	micro
N or n	nano
P or p	pico

Therefore component values can be designated in the way that they are usually specified in circuit diagrams. If you specify none of these multipliers the value will be taken as being in ohms, farads or henrys, depending on which component you are specifying. A multiplier can be used in the form 2.2k or 2k2 as you please. Be careful of the difference between M (Mega) and m (milli); the other letters can be used either lower or upper case without any problems.

Once this information has been entered, the specification of the circuit components is complete, and when the list of components reappears you can press the Esc key to escape from it. Aciran will automatically assume that the highest numbered node is the output node, so that nothing need be done to notify input and output unless you have chosen differently. The next step is to display the circuit layout so as to confirm the component connections.

☐ If at any time you want to get out of an action, press the Esc key.

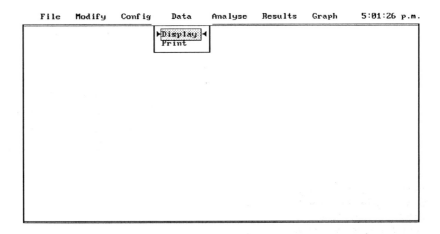

Figure 2.10 *The Data menu option produces the choices of Display or Print, and for checking small circuits Display is faster.*

Displaying the layout is done by selecting *Data* from the main menu, and then *Display*, Figure 2.10. This option is used when you want to review a circuit, and the alternative is to print the data if (and only if) you have a suitable printer connected and switched on. The *Display* option produces its output in the form shown in Figure 2.11, listing the components in numbered order of entry with reference letters and numbers, values, types and node connections. You can now check, preferably with reference to your original circuit, that the connections and component values are correct. If you need to alter the output node it is done by using the *Modify* option of the main menu, of which more later.

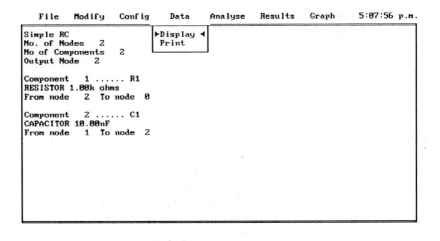

Figure 2.11 *The Display output, showing each component numbered in order of notification, with its value and its node connections. The output node is automatically assigned to the highest number, 2 in this example, but this can be changed manually.*

40 ACIRAN IN ACTION

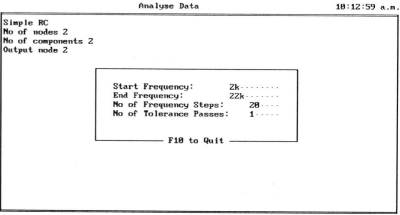

Figure 2.12 *Filling in the Analyse Data form to specify the frequency range and number of analysis steps. Note that for a logarithmic sweep, the ending frequency must be more than ten times the starting frequency.*

☐ You can change component values easily, but not connections. If you later decide that a component should be shifted you have to delete it and add a new component.

The next step is to analyse the data. Select *Analyse* from the main menu to bring up the menu shown in Figure 2.12. This starts with a request for the starting frequency of the range that you want to investigate. This requires some cunning and a rough idea of the circuit action, because it would be ridiculous to ask for a very low starting frequency or to end with a very high one. In addition the ending frequency for a logarithmic frequency range must be more than ten times the starting frequency. In this example, using 2k (2000 Hz) to 22k (22000 Hz) seems reasonable.

You are also asked to type in the number of frequency steps. The larger the number of steps the more precise the graph that will eventually be drawn, but the longer the analysis will take. A figure of 20 is reasonable, and for a circuit which does not produce any violent changes in response a smaller figure would be acceptable. Larger numbers of frequency points are needed for a step or notch filter which would have a steep side. You can ignore the number of tolerance passes in this example because we are not taking tolerance into account.

Once this form has been filled in the analysis will start when you press the F10 key. You will see a report on screen (Figure 2.13) on the progress of the analysis. This repeats the information on start and end frequencies, reminds you that the scales will be logarithmic (suited to a dB graph) and shows the percentage completed, 45% at the time this screenshot was taken. For a simple circuit, this process of analysis is fairly fast, and even for complex programs it will not be a very lengthy process unless a large number of frequency points and tolerance runs are specified.

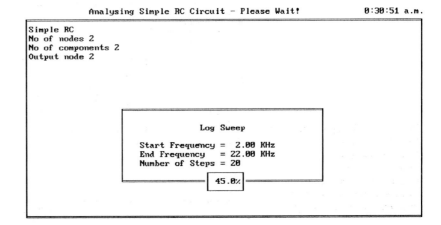

Figure 2.13 *The appearance of the screen when an analysis is in progress; the percentage figure shows how much of the analysis has been completed.*

Transmission Results 10:40:22 a.m.

Frequency(Hz)	Magnitude(db)	Phase(Deg)	Time Delay(Sec)
2.000E+03	-18.084	82.838	
2.255E+03	-17.061	81.937	-9.824E-06
2.542E+03	-16.042	80.926	-9.778E-06
2.866E+03	-15.030	79.793	-9.719E-06
3.231E+03	-14.026	78.525	-9.646E-06
3.642E+03	-13.031	77.110	-9.554E-06
4.106E+03	-12.047	75.533	-9.440E-06
4.629E+03	-11.079	73.782	-9.299E-06
5.219E+03	-10.128	71.845	-9.126E-06
5.884E+03	-9.200	69.711	-8.915E-06
6.633E+03	-8.298	67.375	-8.660E-06
7.478E+03	-7.427	64.833	-8.357E-06
8.431E+03	-6.593	62.089	-8.001E-06
9.505E+03	-5.802	59.155	-7.590E-06
1.072E+04	-5.060	56.049	-7.125E-06
1.208E+04	-4.371	52.801	-6.611E-06
1.362E+04	-3.740	49.447	-6.056E-06
1.535E+04	-3.169	46.030	-5.471E-06
1.731E+04	-2.661	42.598	-4.874E-06

Press Any Key To Continue

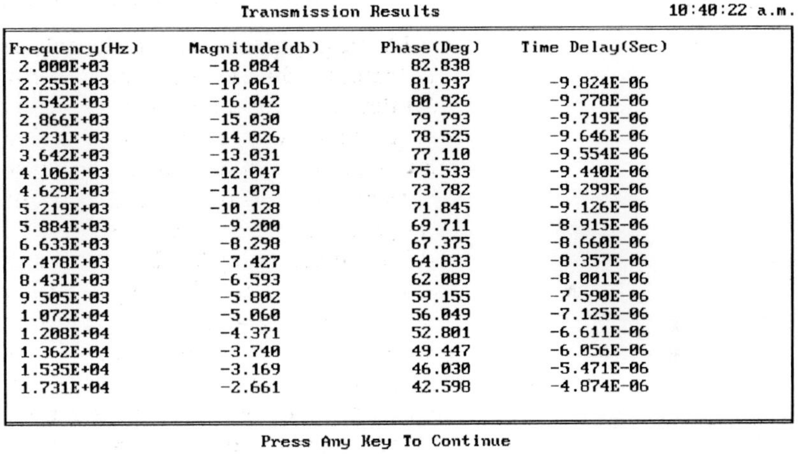

Figure 2.14 *The primary output of Aciran is this table of magnitude and phase angle for the circuit output relative to its input. Only a portion is shown in this screenshot, and the other parts are viewed when any key is pressed.*

Now, at last, we can see some results. When the analysis is complete, the first screen output is a table (Figure 2.14), which usually requires you to press a key to see the second part. You can, of course, print this table or save it to a file. This is done, once the analysis has been completed, by selecting *Results* and choosing one of the three options of *Display*, *File* or *Print*. The *Results* option allows you to see the results of an analysis again without having to repeat the analysis, and using the *File* or *Print* options makes it easy to keep a copy to compare with the results of analysis of a modified circuit.

☐ A file version can be edited and printed using a word processor.

The table shows steps of frequency, amplitude (magnitude), phase and time delay for the circuit. The frequency figures are usually *odd* in the sense that they do not follow a pattern such as 1, 1.5, 2 and so on. These frequencies are obtained by taking the range that you have specified and calculating a set of frequencies that will follow a logarithmic pattern (the spacing is not even, but increases as the frequency increases).

If you use the *Results* menu and opt to make a file of these figures you will be asked for a filename (which can include a drive letter and directory path if needed): see later.

Following analysis and making any copy of the table to file or to printer, you can use the *Graph* items on the menu to see what the response of the circuit looks like. When you select *Graph* (with the *Config* settings used in this example) you will see a set of three graphs, of which the first is the plot of amplitude against frequency, Figure 2.15. This illustrates the steps of frequency and dB values that you can expect in a logarithmic graph. It is not easy on a plot like this to locate the 3 dB point with any precision but it corresponds to the 16 kHz mark when carefully examined. Note that it is very much more difficult to make such readings on a wider scale such as 1k to 100k.

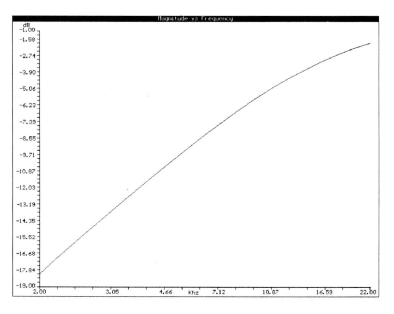

Figure 2.15 *The graph produced by Aciran for the circuit example. The steps of amplitude and frequency are never simple numbers because they have to be calculated by the program.*

☐ For estimating a 3 dB turnover frequency it is usually easier to find the maximum response of the listing, subtract 3 dB and then look for the nearest figure in the listing. This is always likely to be more precise than trying to measure from a graph. Use the graph only as a visual picture of the response.

The following graph, Figure 2.16, is of phase shift plotted against frequency. This shows the phase in degrees plotted against the logarithmic frequency scale and once again, the expected 45° shift occurs at around 16 kHz. The last of the graphs, Figure 2.17, is of time delay through the circuit expressed in microseconds for the usual frequency range.

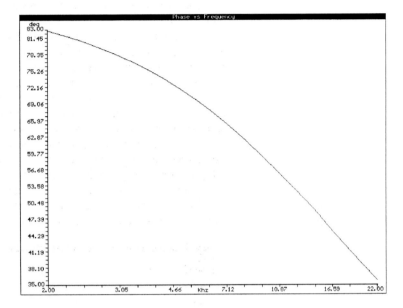

Figure 2.16 *The graph of phase shift plotted against frequency for the circuit example. The units are degrees.*

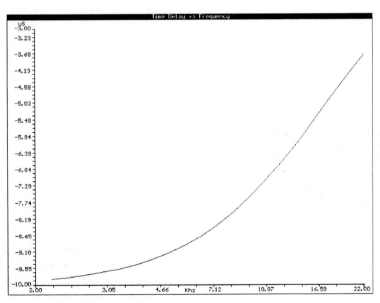

Figure 2.17 *The graph of time delay in microseconds for the circuit example. Note that all three graphs are produced automatically; you cannot opt to produce just one, or a selection.*

We can now summarise the actions for analysing a circuit.

1. Label the circuit diagram with nodes, using 0 for earth, 1 for input and the highest number in the set for output.
2. Run Aciran, and select *New* to type in the components, values and nodes.
3. Check using the *Display* option from the *Data* menu.
4. Select *Analyse* and fill in the frequency range and number of frequency points to get the table of results.
5. Use the *Results* menu to see the table again on screen, to save it to a file or to print it.
6. Use the *Graph* menu to display the graphs of amplitude, phase and time delay all plotted against frequency.

Saving and loading

Once you have used Aciran for analysing a circuit, it is very convenient to save the circuit information so that it can be used again. The circuit files are very short, so that the amount of disk space needed for even a substantial library of circuits is not large. The *Save* action ensures that you can recall the circuit again and, more important, it allows you to use the same circuit with different component values, or the same circuit with some new components added. This considerably reduces the amount of work involved in specifying a circuit to Aciran.

A circuit is saved by using the *File* menu and selecting *Save*. You will be asked to supply a filename, which should include a directory name if your files are being saved to a directory, and you should use something that you will remember later. If you simply use names like CCT1, CCT2 and so on, it becomes very difficult after a few months to remember what each circuit consists of. As it happens, it is easy enough to get a reminder *before* starting Aciran if you have filled in the description for each circuit. This is done when running MS-DOS by using the CD command to get to the directory where the files are stored and using the TYPE command in the form:

```
TYPE CCT1.CIR          (Press ENTER)
```

Most of the display will be of meaningless symbols but the words of your circuit description will appear in plain English.

Nevertheless, it is better if you use filenames such as LOPASS1 or BANDPS2 to remind you without needing this form of investigation. You need supply only the main part of the filename, not the CIR portion (the extension) which follows the dot. The main part of the filename is subject to the usual rules of MS-DOS filenames, and in case you are not familiar with the rules, a summary follows.

To start with, there are names that you cannot and must not use. You cannot use any of the names that appear on the MS-DOS distribution (master) disk, because these are reserved for the programs that bear these names. In addition, there are *internal* commands, stored in memory, whose names you must not use. These are:

BREAK	CD	CHCP	CHDIR	CLS
COPY	CTTY	DATE	DEL	DIR
ECHO	ERASE	EXIT	FILES	FOR
GOTO	IF	MKDIR	MD	PATH
PAUSE	PROMPT	RENAME	REM	RMDIR
RD	SET	SHIFT	TIME	TYPE
VER	VERIFY	VOL		

Any attempt to use these names other than for the programs that they represent may cause problems. If you work, using a floppy drive, with the MS-DOS distribution disk in place, or with a hard disk using a path to the MS-DOS external command programs, then you need to avoid using any of these filenames also. These are:

APPEND	ASSIGN	ATTRIB	BACKUP	CHKDSK
COMMAND	COMP	COUNTRY	DISKCOMP	DISKCOPY
EXE2BIN	FASTOPEN	FDISK	FIND	FORMAT
GRAFTABL	GRAPHICS	JOIN	KEYB	KEYBUK
LABEL	MODE	MORE	NLSFUNC	PRINT
RECOVER	REPLACE	RESTORE	SELECT	SHARE
SORT	SUBST	SYS	TREE	XCOPY

It is most unlikely that you would use any of these names for circuit files, but it is as well to know that using these names could cause problems. Various programs that you use will also have their own prohibitions, all designed to prevent you from damaging essential files on the disks. In general, a suitable filename will consist of up to eight characters, the first of which must be a letter. The other characters can be letters, or you can use the digits 0 to 9, or the symbols:

```
$ # & @ ! % ( ) _ { } ' ~ ^ `
```

The following characters must *not* be used:

```
. + = [ ] ; : , . / ?
```

nor can you use the space, the tab, or the Ctrl character.

Unless it's particularly important to you to use the permitted symbols shown above, they are best avoided, particularly symbols like the single inverted quotes, which are easily confused or overlooked. A good rule is to use words with a digit or pair of digits used at the end to express versions, like LOPASS1, LOPASS2, ..., LOPASS15 and so on. Allowing up to eight characters means that you may have to abbreviate names that you would want to use, like ACTFILT or TONCTRL, but it should be possible to provide a name that conveys to you what the file is all about. Whether you use lower-case or upper-case characters, MS-DOS will convert all filenames to upper-case.

A full filename for a hard disk should include the directory name. When you have been working in the ACIRAN directory, the box that prompts you to enter a filename will start with C:\ACIRAN\ so that you can, if you have followed the advice in this book, type CIRCUITS\ followed by the name of the file. If you are using a floppy disk and have followed the advice in this book you can type A:\CIRCUITS\ followed by the filename for the circuit. If you have loaded a file in from the \CIRCUITS directory and altered it, you will be shown the old name when you want to save it, and you should alter the name (not the directory path) unless you want the older version to be scrapped. If you have loaded in ACIRAN\CIRCUITS\LOPASS1.CIR, then saving another file of this name will destroy the old file, but changing it to ACIRAN\CIRCUITS\LOPASS2.CIR will allow both versions to exist on the disk. Only the last part of the name needs to be changed – you do not need to retype the whole of it.

Once a circuit has been saved, it can be loaded in again, using the *Load* option of the *File* menu, each time you use Aciran. This allows you to work with an ever-growing file of circuits, and since many circuits fall into classes with similar patterns (low-pass, high-pass, etc.) you can often make a new circuit by a few modifications to an existing one, as noted earlier.

Saving and using results

When an analysis has been carried out, and the *Results* menu is used, you have the option of saving the results – meaning the table of results – to a file. Choosing this option will bring up a box:

```
Enter Filename:
```

which will contain a default name that you can edit (backspace over the name to delete it) to suit yourself. You can save to hard disk or floppy disk as you choose.

When the results of an analysis have been saved on to a disk in this way, the file is a plain ASCII one, meaning that it can be read by any word processor or desktop publishing equipment. You can also display the file on the screen, using the TYPE utility that is built into MS-DOS. If, for example, the file is called CCT1.LST and is on a disk in the A: drive then it can be shown on screen by typing:

```
TYPE A:CCT1.LST
```

If this scrolls past too quickly to see you can make it use the screen one 'page' at a time by typing instead:

```
TYPE A:CCT1.LST | MORE
```

provided that the program called MORE.COM is located on the hard disk or the disk in drive A. Another possibility is to print the file by using:

```
          COPY A:CCT1.LST PRN
```

A printer must be connected and ready to use.

Note that any printer can be used for text lists like this – the Epson printer is needed only for graphical printouts.

These files contain a 'header' such as:

```
CCT file on disk TYPE displays

Aciran AC Circuit Analysis V3.0
Copyright (C) 1988,89,90,91   James Herron

Time Stamp :- Tuesday October 15, 1991
Input Circuit File :-    B:CCT2.DOC

Transmission Results for Simple RC
```

This reminds you of the date and the circuit which produced the results and is followed by the complete table.

Log and linear plots

The default plotting method for Aciran is logarithmic, which allows amplitudes to be plotted on a dB scale against a logarithmic scale of frequency values. This is the method that is normally used in text books to display frequency response, but it makes the frequency steps rather more difficult to read. The alternative is to use linear plots, remembering that a linear plot often looks misleading because the range is much more limited. As it happens, the linear plot of Aciran is of frequency only, not of amplitude.

The change is carried out by using the *Config* item on the main menu. Selecting this item opens out a set of options, one of which is *Sweep...Logarithmic*, Figure 2.18. By selecting this and pressing ENTER or clicking the mouse button, you can change *Logarithmic* to *Linear*, allowing you to make linear sweeps. This does not mean that the scales of frequency or amplitude will look any easier to use. The amplitude scale will still be in decibels, and the frequency scale will still be in odd units.

☐ One important advantage of a linear plot, however, is that the ending frequency can be much closer to the starting frequency. This makes it much easier to read where a 3 dB point occurs, or to follow a sharply changing graph characteristic. Note that a logarithmic sweep will be restored the next time you start to use Aciran, but your selection of *Linear* will remain in force until you alter it or leave the program.

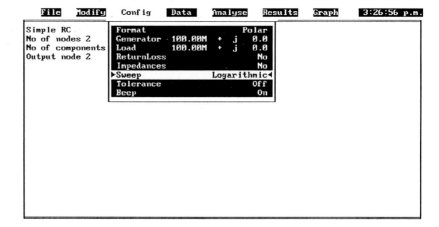

Figure 2.18 *Using the Config menu to change from a logarithmic sweep to a linear sweep by selecting the Sweep line. The item will change alternately each time the line is selected.*

Figure 2.19 shows a small-range graph plot for the RC circuit which has been the subject of this entire chapter so far. The frequency range is 15 kHz to 18 kHz and the choice of range has automatically resulted in the marked frequency points on the graph being 530 Hz apart with the decibel scale using intervals of 0.13 dB. The use of a linear frequency scale over a small range makes it easier to look for the 3 dB point, but it also makes the rate of change of amplitude less easy to read.

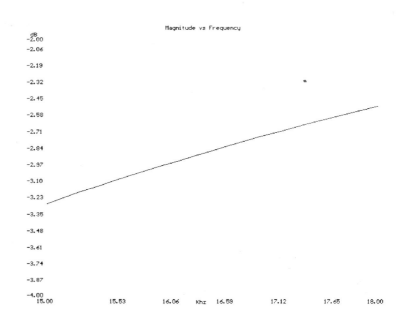

Figure 2.19 *A small range plot for 15 kHz to 18 kHz made by specifying a linear sweep and then this frequency range.*

☐ Remember that the graphs illustrated here are as they appear on the screen – a directly printed graph has a much better appearance because the resolution of the printer can be as high as 300 dots per inch; the screen permits only 75 dots per inch.

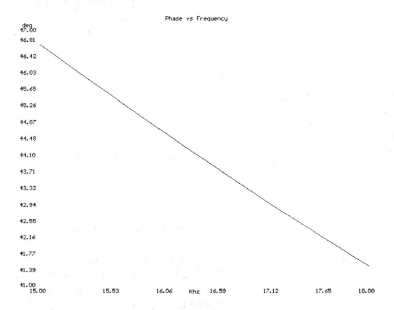

Figure 2.20 *The linear phase shift graph for the same range. This makes it very easy to find the frequency for which the phase angle is 45°.*

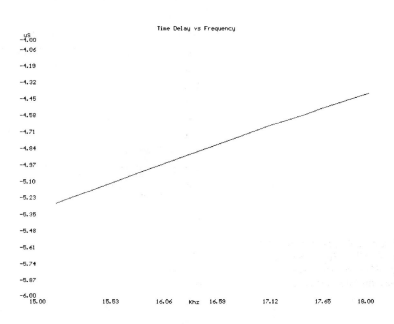

Figure 2.21 *The linear plot of time delay is itself almost a straight line over this range.*

The corresponding linear plot of phase shift is shown in Figure 2.20 above. This allows for easy measurement to locate the frequency for which the phase shift is 45°. For more complex filters in which there is a large phase change around a critical frequency this type of plot can be very useful. The linear plot of time delay against frequency (Figure 2.21) is itself almost a straight line.

☐ Remember that any of these graphs can be printed if you have a suitable printer connected and ready. While a graph is being displayed on the screen, it can be printed by pressing the H key, or, if the computer uses a CGA type of screen and the GRAPHICS program has been run, by pressing the PrintScreen key (sometimes along with the SHIFT key – check with the manuals for your computer and printer).

Editing the circuit

The main point about saving a circuit that has been analysed is that its parameters can be edited. The simplest form of editing is the replacement of one component value by another, and such work starts with loading in the circuit, using the *Load* option of the *File* menu of Aciran. Once the circuit is loaded you will see the summary for the circuit appear, with the title, number of nodes, number of components and output node printed on the screen. You can then select *Modify* from the main menu, and from the list of options pick *Change* (Figure 2.22) allowing you to change the value of a component.

☐ Note that the *Replace* option in this list is a more specialised one. It allows a component which has been deleted to be replaced by another component which will be used in the circuit, and is intended to be used when you are considerably modifying a circuit, not simply changing values.

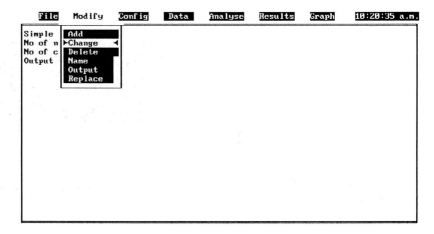

Figure 2.22 *The Modify list, showing the Change option selected.*

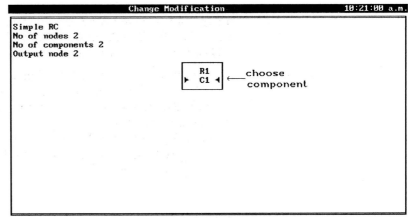

Figure 2.23 *The components list for the Change option, allowing you to select one or more components to change. Note that you cannot change node wiring in this way.*

Taking the *Change* option brings up a list (Figure 2.23), of the components in the circuit. This simple circuit consists of only two components, C_1 and R_1, so that only these appear, but for a complex circuit you might need to scroll down such a list to find the component you wanted to change. In this example, C_1 is selected to be changed by moving the marker from the first component in the list, R_1. The use of circuit descriptions such as C_1 and R_1 is important for this action.

This in turn brings up the specification for C_1, Figure 2.24. You can alter either the value or tolerance here, but the change that is shown

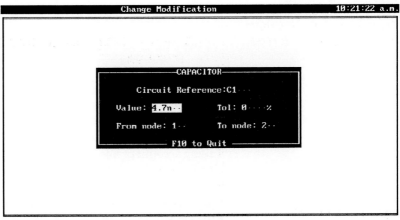

Figure 2.24 *Altering the capacitor value to 4n7 by filling in the Change form.*

being made is the alteration in value to 4.7 nF. Pressing the F10 key will confirm the alteration.

☐ Note that you need to erase the old value before entering a new one, otherwise the new entry will simply follow the old one, making an impossible or bad value. Typing `4.7` on the `10 nf` for example would result in `10n4.7n` which would be rejected.

Once you press F10, the new value will be used and you can analyse the circuit again to produce the table of amplitude, phase and time-delay values over your selected frequency range, logarithmic or linear. Figure 2.25 shows the amplitude plot over the frequency range of 10 kHz to 101 kHz, showing that the 3 dB point is around the expected position – for a precise value you could use a linear plot between 30 and 40 kHz.

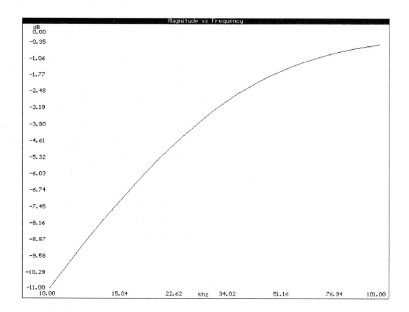

Figure 2.25 *The amplitude frequency plot for the amended circuit, using a range of 10 kHz to 101 kHz and a logarithmic sweep.*

Dealing with tolerances

Taking account of tolerances is by far the most difficult part of any linear circuit analysis. Even for a simple circuit it will involve finding the maximum and minimum values that can be expected for the given tolerance limits and calculating the response of the circuit for these extreme values, so that the graphical plot becomes a band rather than a simple line.

Using tolerances with Aciran is considerably simpler, though it takes longer to obtain an analysis because of the larger number of calculations that have to be done. This is necessary because Aciran is unable to predict as a human operator could which combinations of tolerances have the greatest impact on the results.

☐ If your computer is fitted with a maths co-processor such as the Intel or IIT 80287 (for an 80286 computer), all calculations are very considerably aided – a factor of 50 for speed increase is possible for some types of calculations. The slower machines using the 8088 or 8086 chip will benefit most from the use of a suitable (8087) co-processor.

By default, the use of tolerances should be switched on in the *Config* menu when Aciran is started up, but if you have previously switched this option off during a session with Aciran you will need to restore it before you can use tolerances in the same session. You will also have to switch *Tolerance* on if you have loaded a circuit which was saved with *Tolerance* switched off, because such details are saved along with the other data in the file. Assuming that the use of *Tolerance* is enabled, you can alter the circuit which has been used as an example so as to account for tolerances.

Aciran deals with tolerances by using a Monte Carlo method, something that is not likely to be familiar to every designer of linear circuits. Basically and briefly, this simulates the effect of tolerance by making many sets of figures with the values of the components varied by a random amount (up to the bounds set by tolerance figures) in each run. The tables that are produced in this way are duplicated, one set showing the upper limits that are expected, the other showing the lower limits.

It is at this point that you run up against the conflict between speed and precision. If you want your analysis to reflect the effect of tolerances really precisely, you will need to make a large number of these Monte Carlo runs, but this will take a considerable time, many hours for a large and elaborate circuit. Using only three Monte Carlo runs will produce results reasonably quickly, but though they will indicate the effects of tolerance they will not show the effects so precisely as when a large number of runs have been used.

To show the effects of tolerances on the example, load in the original file and select *Modify*. Alter the Capacitor entry form so as to change to a tolerance figure of 20%, and alter the Resistor entry to make its tolerance 10%. Figure 2.26 shows a composite screen view of both actions. Once these tolerance entries have been made, and the *Config* menu allows tolerance action, any analysis will be able to carry out the tolerance runs.

54 ACIRAN IN ACTION

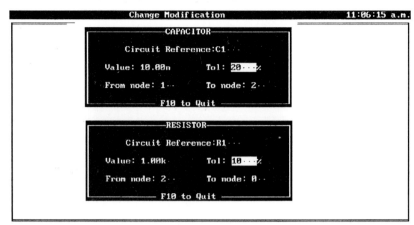

Figure 2.26 *A composite view of the alterations to provide tolerance data. Tolerance must also be switched on in the Config menu, and you must specify a number of tolerance runs in the Analyse menu.*

First, however, the number of tolerance passes must be specified in the *Analyse* form, Figure 2.27. In this screenshot it has been shown as 10, which is a good compromise between speed and precision – a figure of up to 32767 can be used (if you can wait for several days). In subsequent screenshots a figure of 100 has been used.

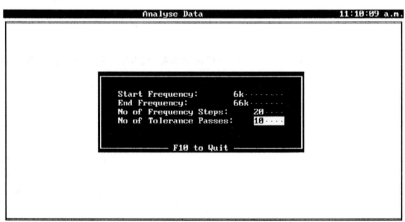

Figure 2.27 *Specifying ten tolerance passes in the Analyse menu, so that a complete circuit analysis will be carried out ten times with values that are varied within the tolerance limits.*

Figure 2.28 shows the analysis in progress with 100 tolerance runs specified. The screen report confirms the logarithmic sweep between 6 kHz and 66 kHz, using 20 steps of frequency; on the foot of the screen, the 'Monte Carlo Run 73 of 100' shows that 100 tolerance runs have been requested and number 73 is in progress – remember that each run is a full analysis and takes the same time as a single run with no tolerance figure.

The result is two sets of tables which reflect the extremes of the tolerance limits. In the screen shot of Figure 2.29, the start of each table set has been combined so that you can see how different the two tables

Analysing Simple RC Circuit - Please Wait! 1:11:25 a.m.

Log Sweep

Start Frequency = 6.00 KHz
End Frequency = 66.00 KHz
Number of Steps = 20

72.6%

Monte Carlo Run 73 of 100

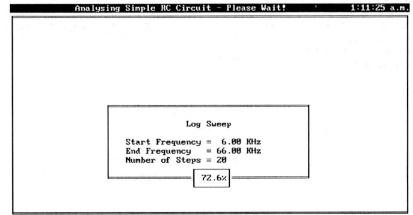

Figure 2.28 *Tolerance runs being made when 100 runs have been specified. Thepercentage figure shows how much of the work has been done, and the number of the Monte Carlo run is also shown.*

Transmission Results 11:20:37 a.m.

Frequency(Hz)	Magnitude(db)	Phase(Deg)	Time Delay(Sec)
6.000E+03	-7.059	74.313	
6.764E+03	-6.605	72.772	2.364E-05
7.626E+03	-5.618	70.446	1.802E-05
8.597E+03	-5.123	68.490	1.835E-05
9.692E+03	-4.199	65.397	1.841E-05
1.093E+04	-3.740	63.127	1.744E-05
1.232E+04	-3.244	59.909	1.550E-05
1.389E+04	-2.582	55.817	1.590E-05
1.566E+04	-2.086	51.780	1.168E-05

Frequency(Hz)	Magnitude(db)	Phase(Deg)	Time Delay(Sec)
6.000E+03	-11.360	63.662	
6.764E+03	-10.569	62.130	-3.510E-05
7.626E+03	-9.507	58.418	-3.457E-05
8.597E+03	-8.715	56.330	-3.094E-05
9.692E+03	-7.611	51.929	-3.915E-05
1.093E+04	-6.897	49.451	-3.016E-05
1.232E+04	-5.997	46.500	-2.760E-05
1.389E+04	-5.008	42.026	-2.865E-05
1.566E+04	-4.171	38.141	-2.047E-05

Press Any Key To Continue

Figure 2.29 *Two views of the output tables when tolerance runs have been carried out. These figures show the extremes of amplitude and phase response for the selected tolerances.*

are: by around 4.3 dB of amplitude at the lowest frequency, less so at higher frequencies. The phase shift figures are also markedly different, by around 10.5° at 6 kHz and more than 13.5° at 15.66 kHz.

This leads also to two sets of graphs. The amplitude graphs (Figure 2.30) show the spread of values – note that the graph line shape is more irregular because even with 100 Monte Carlo runs the spread of values is not enough to make for completely smooth lines. It is often useful to superimpose the graph obtained with zero tolerance over the graphs

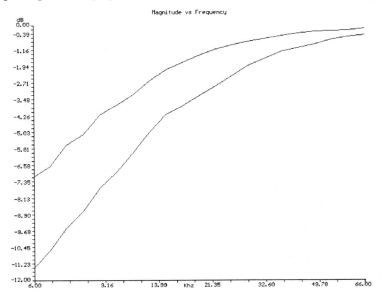

Figure 2.30 *The twin-graph produced when tolerance runs have been used; note that obtaining this graph is not straightforward in Version 3.0.*

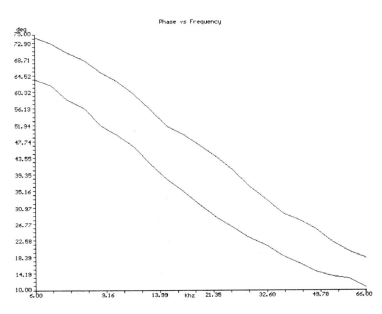

Figure 2.31 *The phase-shift graphs for tolerance runs, showing the expected spread of this quantity.*

DEALING WITH TOLERANCES   57

obtained with tolerance values. The phase graphs, Figure 2.31, also reflect the considerable spread of values and the time delay graphs of Figure 2.32 look odd until you check back with the tables to find that one set of figures is taken as positive and the other as negative.

Now look at the amplitude graphs that result when only three Monte Carlo runs are used (Figure 2.33). These are decidedly rough and jagged, and would not be a very good basis for predicting the effects of

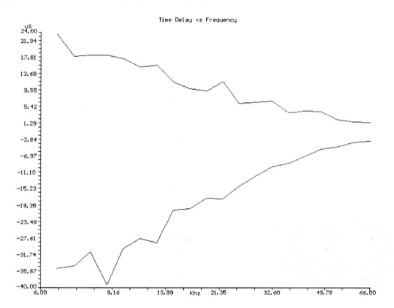

Figure 2.32 *The time-delay graphs. The graphs are asymmetrical because of the random way in which tolerance values are calculated, and would be smooth only for an infinite number of runs.*

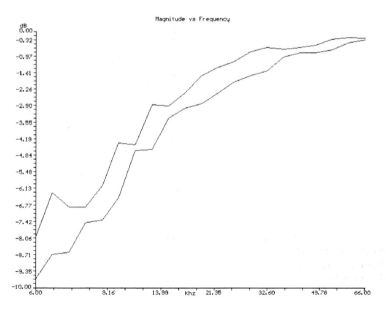

Figure 2.33 *The very jagged graph which is the result of using too few Monte Carlo runs for tolerance.*

tolerances. The best approach, then, is to make a run using zero tolerances so as to determine the ideal analysis, and then to make a set of tolerance runs, using as many as time permits; since it is machine time rather than human time, there is no reason why a large number should not be used.

☐ A peculiarity of Aciran Version 3.0 is that the twin-graphs display does not appear at once. If you find that after seeing the two sets of tables you cannot obtain the twin graphs, try turning back to the Analysis menu and pressing the Esc key to escape from the menu rather than the F10 key to carry out another analysis. Now select *Graph*, and the dual-graph display should appear. Later versions do not exhibit this problem.

Frequency insensitive circuits

The circuits dealt with so far are all those which have a response that depends on frequency, and most of the applications of Aciran will be for the analysis of such circuits. There are, however, circuits such as attenuating pads which are intended to provide a response that is independent of frequency, and Aciran can be used to find what attenuation such a pad produces. More important, Aciran can check that the correct impedance levels are being used for an attenuator, and can also find (given realistic values of stray capacitance) for what frequency range the assumptions of fixed attenuation will hold good. Aciran can also show how the attenuation of a circuit will be affected by component tolerances.

There are three basic types of attenuator pads – L, T and Pi – and the T and Pi types can exist in symmetrical or asymmetrical form. It is not the purpose of this book to go over attenuator pad theory, but since Aciran is so useful for such circuits, some detail of what the circuits do and how they are designed may be useful, as it avoids the need to refer to other texts until you need more information.

The L type of attenuator (which is more of a Greek gamma shape) is used when signals have to be passed from one impedance to another and attenuated. Referring to the diagram of Figure 2.34, the values of R_1 and R_2 can be found if the input impedance Z_{in} and output impedance Z_{out} are known, and the equations are:

$$R_1 = \frac{Z_{in} Z_{out}}{R_2}$$

$$R_2 = \sqrt{\frac{Z_{in} Z_{out}}{Z_{in} - Z_{out}}}$$

$$\text{loss} = 20 \log \left(\frac{Z_{out}(R_1 + R_2) + R_1 R_2}{R_2 Z_{out}} \right)$$

The resistors are as indicated in Figure 2.34. The value of Z_{in} is assumed to be higher than that of Z_{out}.

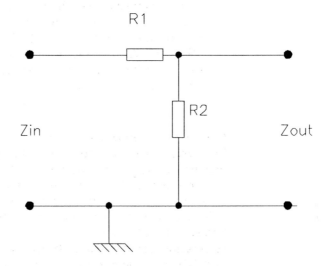

Figure 2.34 *The basic form of L attenuator pad.*

For example, if the input impedance is 300 ohms and the output impedance is 75 ohms, then an attenuator pad design has to start with the calculation of R_2. In this example, R_2 is given by:

$$\sqrt{\frac{300 \times 75^2}{300 - 75}}$$

which is 86.6 ohms. R_1 is found from this as:

$$\frac{300 \times 75}{86.6}$$

This gives $R_1 = 259.8$ ohms. These figures show one disadvantage of this and most other attenuator pad designs – the values are not exactly easy to pick out of a box. The voltage attenuation of this pad should be around 17.5 dB.

☐ Some formulae quoted in reference books for these pads do not give the correct values of attenuation.

We can now analyse the circuit with Aciran, using two nodes and the resistor values shown. When the values have been entered, use the *Config* menu to type in a value of 300 for the source impedance and 75 for the load impedance. Also in the *Config* menu, switch on *Impedances* so that Aciran can report on the actual impedance found both at input and at output. The results of the analysis are as expected: input impedances of 300R and 75R respectively, and attenuation of –17.459 dB.

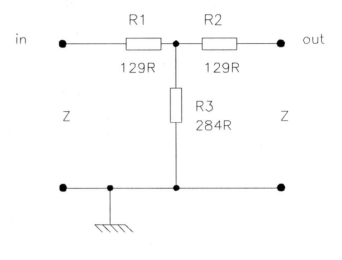

Figure 2.35 *The basic form of T attenuator pad.*

The T-pad circuit, Figure 2.35, is intended to be used between identical impedances and to produce a fixed value of attenuation. The design of a T-pad is not so simple as that of the L-pad, and you need to start by finding the power ratio N from the desired attenuation in dB. From there, you can find values for the resistors in terms of N. The equations are:

$$N = 10^{A/10}$$

$$R_1 = R_2 = Z\,\frac{\sqrt{N}-1}{\sqrt{N}+1}$$

$$R_3 = \frac{2Z\sqrt{N}}{N-1}$$

We can illustrate their use for the attenuator in the drawing, which requires an attenuation of 8 dB between 300R resistance levels.

The value of N is found by using a calculator to work out ten to the power 0.8, and this value is found to be 6.3095. When this is inserted into the formulae for R_1, R_2 and R_3, the values are 129.147 and 283.846, so that practical values of 129 and 284 can be used. Later we can look at the effect of using components of the nearest preferred values. When Aciran is used for analysis, the attenuation is found to be –7.992 dB, with an impedance level at input and output of 300R, checking once again that Aciran delivers the correct outputs for such circuits.

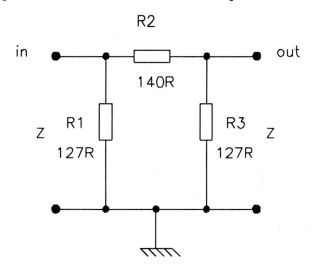

Figure 2.36 *The basic form of Pi attenuator pad.*

$$Z = 75R$$

The last remaining form of attenuator is the Pi-pad, Figure 2.36. The power ratio N is used to calculate the values for this pad as well, and the resistor values are found from:

$$R_1 = R_3 = Z \frac{N-1}{2\sqrt{N}}$$

$$R_2 = Z \frac{\sqrt{N}+1}{\sqrt{N}-1}$$

In the example, a 75 ohm line requires a 12 dB attenuator, so that the value of N is found from 10 to the power 1.2, and is 15.849. Working out the resistor values from this gives 139.87 for R_2 and 126.61 for R_1 and R_3. Values of 140 and 127 ohms respectively have been used for Aciran. The results are attenuation of –11.974 dB and impedance of 75.66 ohms, once again confirming that the analysis follows established formulae.

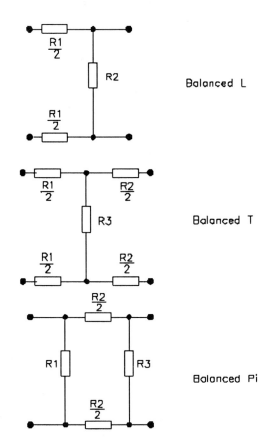

Figure 2.37 *The balanced forms of the same attenuator pads.*

These attenuator circuits can be used either single-ended, as shown here, or balanced. Figure 2.37 shows the balanced forms in which the resistor values use the reference numbers established for the single-ended form. The balanced forms are used when the lines on each side of the attenuator are balanced twin lines and, as the drawing shows, the series resistors are all halved in value as compared to the unbalanced form, with the parallel resistor values unchanged. The balanced form of the Pi-pad, sometimes called the O-pad, is preferred because it uses four resistors rather than the five of the balanced T-pad (or H-pad) – the H-pad would be used only if the resistor values were particularly convenient.

It is also possible to design asymmetrical attenuators in which the resistor values are all different. For example, whereas a symmetrical T attenuator has two resistors in series with the signal and of equal values, an asymmetrical T attenuator would use different values for these resistors. Asymmetrical attenuators can sometimes make it possible to use values closer to preferred values. Such attenuators are less often used than the symmetrical type, and can be analysed by Aciran in the same way.

Effect of stray capacitance

A valuable contribution that Aciran can make to attenuators is to analyse the effect of strays on pads that have to operate at high frequencies. In general, attenuator pads are used for frequencies that can range from audio to UHF, so that their use at the higher frequencies is always likely to produce different amounts of attenuation unless the stray capacitances can be kept at very low values.

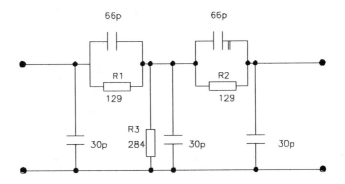

Figure 2.38 *A T-pad 8 dB attenuator for a 300 ohm line, with stray capacitances added.*

Figure 2.38 shows a T pad attenuator for a 300 ohm line with stray capacitances added. These stray values are on the high side, but their relative values are reasonable. They cause a roll off from the normal 8 dB attenuation of this pad starting at around 7 MHz, illustrated in the Aciran graph of Figure 2.39. The input impedance and output impedance graphs also start to dip at the same turnover frequency, ending up with a value of 35 ohms at 100 MHz. In this (admittedly exaggerated) example, then, the attenuation is not maintained to VHF frequencies.

A pad is seldom so poor as this, and with suitable construction stray capacitances can be kept to much lower levels. The principles, however, are important because they point to the attenuation ratio of such a pad becoming considerably greater at the higher frequencies. The answer is frequency compensation, which in this case means adding capacitance deliberately to the series arms.

The calculation is relatively simple – the product CR should be made to be the same in each arm. If we take it that the parallel stray capacitance is 30 pF on the parallel arm of 284 ohms, then the product for this arm is 30 x 284, and this ought to be the same as the series arm resistance multiplied by its total capacitance, 129 x C. This makes C equal to 66 pF, and if this value is substituted in place of 20p, using the *Modify* menu and then the *Change* option, the pad should provide the

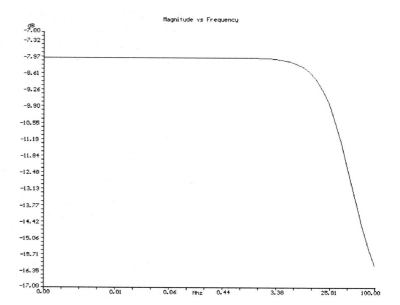

Figure 2.39 *The response of the T-pad with stray capacitances as shown.*

correct ratio again. The effects of this, equivalent to adding 46p capacitors across the series arms, are shown in Figure 2.40.

☐ Compensation like this can be used to make the attenuation value correct but it does not correct the phase changes, nor does it correct the input and output impedances of the pad at the high frequencies.

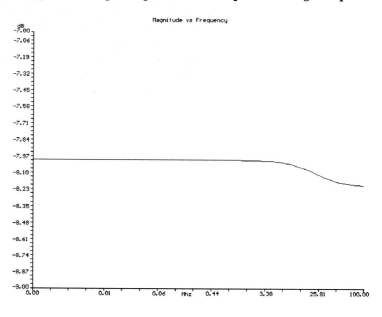

Figure 2.40 *A compensated pad, in which the effects of stray capacitances are balanced by adding capacitors. Though the compensation is not perfect, the attenuation is maintained to a much higher frequency.*

Attenuator pads should be constructed on insulating boards with a single point earth rather than a strip or sheet, and with resistors arranged so that the stray capacitance will be minimal. Stray values as low as 2 pF can be obtained with good construction, though any metal casing or shielding will raise the stray values.

Effects of tolerances

Attenuator pads require resistor values that almost always do not conform to the preferred value ranges, and Aciran can be very useful in showing what the effect will be of using preferred values, and of the tolerance of values. In the following examples, we will use the Pi pad of Figure 2.36, whose calculated values were 139.87 ohms and 126.61 ohms. Suppose we use values of 150R and 120R in this pad and analyse with Aciran, using *Config* with 75 ohm terminations and turning *Impedance* measurements on.

The analysis gives an attenuation of –12.568 dB, not too far from the 12 dB of the original specification. The impedance is given as 74.4 ohms, again not too far out. Unless very precise attenuations are needed, then, it is perfectly possible to make use of preferred values. The next step is to see how the use of 5% tolerances would affect these readings.

The tables that are printed (not illustrated here) show the variation in attenuation caused by assuming 5% tolerances on the preferred values of 150R and 120R. The attenuation range is from –12.127 to –12.990 and the impedance levels range from 71.0 ohms to 77.7. The possibility of the attenuation being almost 1 dB out would make this unacceptable for many purposes, and the possible variation in impedance could also cause problems. In general, it is the tolerance of preferred value resistors that causes problems, and 1% or closer should always be used for constructing attenuators. Where possible, values that are not preferred values should be constructed by adding preferred values in series (reducing the total stray capacitance) rather than in parallel (which would increase strays).

3 Other passive filter circuits

Using examples such as a simple RC single-stage filter or an attenuator pad has the advantage of familiarity, so that you can check that the results of Aciran analysis correspond to what you already know about these circuits. You can also deal with such circuits in much more detail than you would if everything had to be calculated by hand, taking into account matching impedances, tolerances, stray capacitances and all the rest of the real-world problems that are almost always conveniently ignored in texts. The real advantages of Aciran, however, are found when you start to work on a circuit which is not so familiar or so easy to analyse. Consider, for example, the circuit of Figure 3.1, which is a ladder network containing resistors and capacitors only. This, like all ladder networks, takes quite a long time to analyse by conventional methods, and is a useful circuit to illustrate the effects of assuming terminating impedance values. The behaviour of the circuit can change considerably when different terminating values are used, and conventional analysis is too tedious to take this into account.

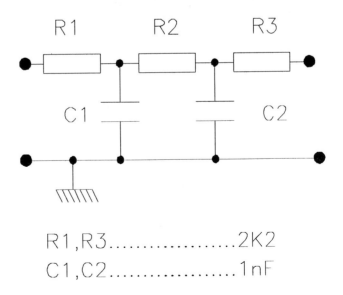

Figure 3.1 *A simple ladder low-pass filter consisting of resistors and capacitors only.*

One conventional method of analysing this type of network would be to find the equivalent resistance of R_3 and the terminating resistance, and redraw the circuit with this resistance across C_2. The impedance of this resistance along with C_2 could be calculated, and the attenuation caused by R_2 and this load also. The equivalent load across C_1 could then be calculated, leading to another attenuation figure, and the attenuation figures would be combined to find the total attenuation. Remember that all of these calculations would use complex numbers so that each result contains a real (in-phase) and an imaginary (90° phase) part. A more realistic alternative is to look up the formula for a ladder network and insert values – but many texts quote the formula in a very confusing way, with little explanation of the meaning of symbols. This makes such a circuit ideal for the application of Aciran.

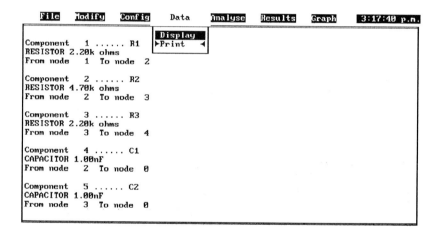

Figure 3.2 *The list of components and nodes. Always check this list before analysing to ensure that the circuit is correctly specified.*

When the circuit is drawn with its nodes numbered and the component values inserted, these quantities can be typed into Aciran and the results are as shown in Figure 3.2, listing the component references, values and nodes. At this point you should always check that the values and positions of the components are correct because if you enter the components from a circuit diagram it is quite easy to miss out a component or mistype a node number. Aciran allows you to correct the value or tolerance of a component, but not its nodes, and if you mistype a node number you have to delete the component and try again.

☐ Note that when a component is deleted in this way, its reference number cannot be used by another component. This allows you to revive the deleted component, but you have to remember this point when you look at Aciran listings.

A quick preliminary try-out on this example using a wide frequency range suggests that a suitable range for a logarithmic analysis would be 5 kHz to 55 kHz (remember that Aciran needs a range greater than 10:1 for a logarithmic plot), and the graphical result is shown in Figure 3.3, showing a sloped characteristic which is 10 dB down at 55 kHz. This, remember, is for zero tolerance and with 100 M terminating impedances. One of the considerable advantages of using Aciran is that you can make preliminary runs like this with a very wide frequency range in order to find in advance what range should be specified for more precise analysis.

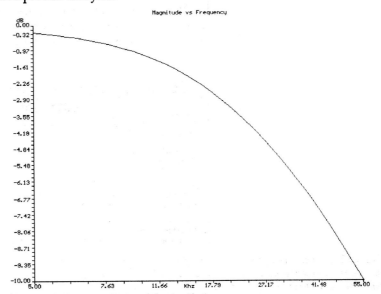

Figure 3.3 *The amplitude characteristic for the circuit with infinite terminating impedances assumed.*

Now look at the effect of making the impedance levels equal to 1k at each side. This is done by using the *Config* menu and selecting *Generator* and *Load* in turn. For each of these you are asked to specify a real part (for which you answer 1K) and imaginary part, for which you enter nothing (unless you want to simulate the action of capacitive or inductive impedances). The results of retyping the impedance values appear as in Figure 3.4 and the resulting graph is illustrated in Figure 3.5. The differences caused by using the output impedance of 1k are enormous, including a 20 dB amplitude drop at low frequencies, even at the starting frequency, and a change of only 3 dB from this at around 55 kHz. The original response, which provided a cut of 10 dB at 55 kHz and virtually none at low frequencies, has been replaced with one that has a 20.13 dB cut at 5 kHz and about 23.5 dB at 55 kHz. If you have no idea what terminating impedances should be used, Aciran will show the input and output impedance of the circuit for the range of frequencies; this aspect of Aciran analysis will be explained in greater detail later. No simple network containing reactive components is ever likely

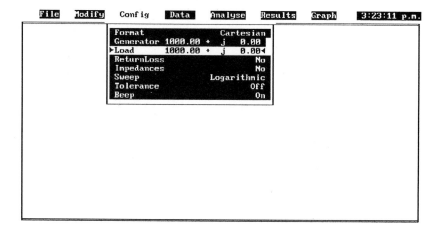

Figure 3.4 *The new terminating values typed, using the Config menu.*

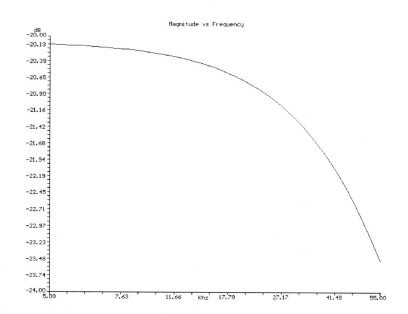

Figure 3.5 *The amplitude graph following analysis, showing the drastic changes due to using lower impedances.*

to have a truly constant value of input or output impedance, so that terminating values are often a matter of 'best-fit' rather than of precise value.

More elaborate RC ladder networks can be analysed just as easily. Figure 3.6 shows a circuit that contains ten components and a total of eight nodes. Each series arm in the ladder network consists of a capacitor, but the parallel arms each contain a series capacitor and resistor combination. This would be a considerable task to analyse from scratch, but is no problem when Aciran is used. On inspection, this

OTHER PASSIVE FILTER CIRCUITS 71

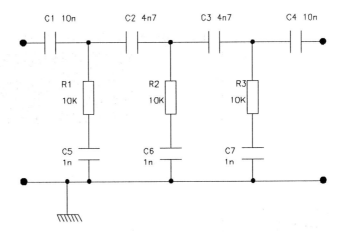

Figure 3.6 *A much more elaborate ladder network, using seven capacitors and three resistors.*

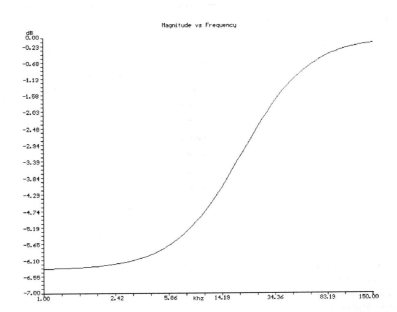

Figure 3.7 *The amplitude graph for the ladder circuit, showing its S-shaped response.*

would be expected to be a high-pass filter with a limited attenuation for the lower frequencies, and when Aciran is used the S-shaped graph of Figure 3.7 shows that this is indeed what happens, with the attenuation at 1 kHz limited to around 6 dB and with a rising response centred around 22 kHz. The phase response is also interesting (Figure 3.8) with a distinct peak of 20° at around the 22 kHz region.

Now look at the effect of modifying this so that the capacitors C_5, C_6 and C_7 are changed to 10 nF, and the analysis is now run for the range of 100 Hz to 150 kHz. The amplitude graph, Figure 3.9, shows a greater attenuation of 29 dB at the lowest frequencies with the central part of

72 OTHER PASSIVE FILTER CIRCUITS

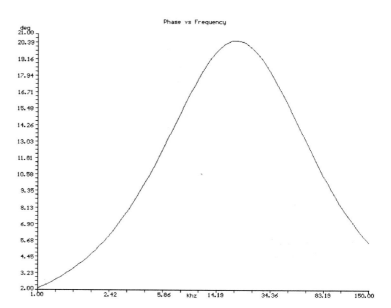

Figure 3.8 *The phase response for the circuit exhibits a peak value of 22° at about 22 kHz.*

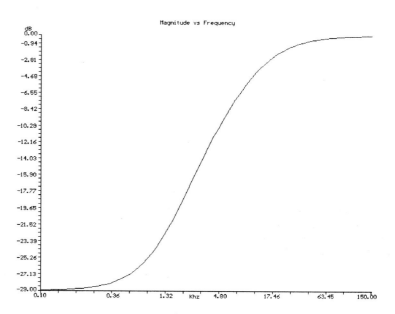

Figure 3.9 *The amplitude graph obtained when the capacitor values are changed as described.*

the rising portion now positioned at a much lower frequency of around 4 kHz. The phase graph of Figure 3.10 shows the peak phase shift of about 80° at around 4 kHz and the time delay graph, not illustrated here, shows zero delay at about 3640 Hz.

With the capacitors C_5 to C_7 held as they are, changing the resistors to 2k2 causes the amplitude graph to change to that shown in Figure

OTHER PASSIVE FILTER CIRCUITS 73

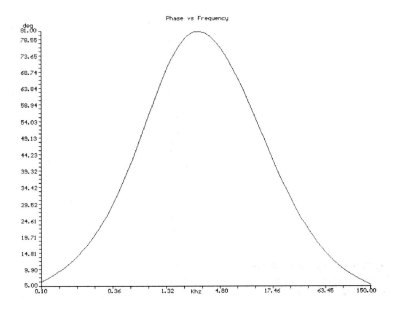

Figure 3.10 *The phase-shift graph now has a much more prominent peak.*

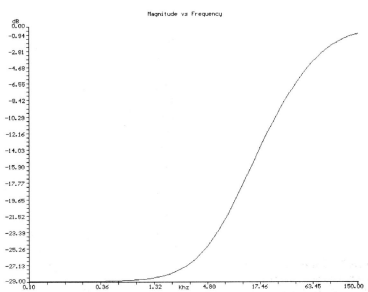

Figure 3.11 *The effect now of changing the resistor values is to shift the position of the rising portion of the amplitude graph.*

3.11. The shape of the graph is very much the same, but the cross-over frequency has now been considerably raised. The attenuation is a maximum of 29 dB below about 1 kHz and the rising portion is now centred about 17 kHz. The phase graph confirms this (Figure 3.12), with a peak of around 80° again at the same cross-over frequency, which is also the frequency of minimum delay.

74 OTHER PASSIVE FILTER CIRCUITS

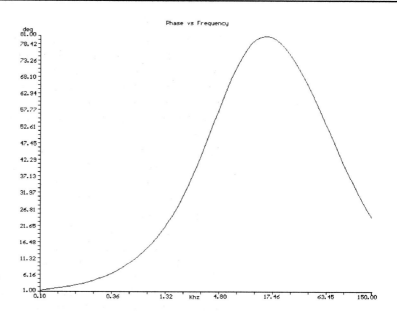

Figure 3.12 *The phase graph for the circuit with reduced resistor values.*

All of these graphs are of the same fundamental shape, which is determined by the type of circuit rather than by the values that are used, and one of the very valuable features of Aciran is that it allows you to develop a feeling for what a circuit can do without the need for endless computation or construction and measurement. In this respect, Aciran is both a very valuable development tool and an excellent aid for students of linear circuitry.

☐ You can, for example, develop a filter circuit for yourself without too much attention to theory, provided you have an example of the circuit with marked components. After altering the time constant values you can use Aciran to check that the response is now in the region(s) that you want.

The range of filter circuits using R and C or R and L is truly enormous, and only a few examples are being considered here. Such filters have a relatively poor action when used in a purely passive way, and come into their own only when used as active filters: see Chapter 5. Active filters using R and C only are to a large extent replacing older types of filter designs based on LC circuits, because the active filter dispenses with inductors and is much easier to make in integrated form. It is also free of the side effects of LC filters that arise from resonances between the inductors and stray capacitances, or because of stray coupling between inductors. The tendency in circuit design over the last 20 years has been to eliminate inductor use as far as possible, using strip line tuning for UHF, transfilters for IF, and active filters for lower frequencies.

Analysing LC filters

The analysis of LC filters is something that traditionally depends heavily on the use of well-documented examples. It is often very difficult, following the normal paths, to see what would be the effect of such items as stray capacitance, winding resistance and other factors that are not taken into account in the design of filter circuits. The calculations required even for comparatively modest LC filters almost immediately start to become tedious and like anything else that is tedious and repetitious, prone to error. In this part of the book, we shall look at some standard LC filter designs to assess how much easier it is to use Aciran than to follow the text-book calculations, but not all calculation is avoidable if you are designing circuits. If, however, you simply want to modify a circuit for your own use, Aciran can show if your modifications are suitable.

This requires a circuit diagram with component values for the filter you want to use, and you need to have some idea in particular about the resonant frequencies of parallel and series LC components. To alter values for another frequency range will require changes to all of the L and C values, making the resonances occur in the new frequency range, but maintaining the same ratios of L to C values, because these often determine the characteristic impedance values. The best method is to make your calculations on the basis of LC values first, as follows:

1   For each resonant LC pair, find a new LC value for the new resonant frequency, using the equation:

   old frequency x square root(old LC) = new frequency x square root(new LC)

2   With the new LC values calculated, find values of L and C which maintain the same ratio of L/C as in the original.

All LC filters will have a characteristic impedance value and should be used with this value, which is always a compromise value, terminating them on each side. For the simpler filters it is relatively easy to calculate what this terminating resistance ought to be, but for the more complex filters it is often impossible to find a suitable terminating resistor. This makes the use of Aciran, allowing you to experiment with terminating resistor values, even more valuable. Note that the use of Aciran with typical terminating resistor values such as 600R will often make nonsense of some theoretical filter response diagrams.

In general, when Aciran is used with infinite terminating impedances, it is possible to obtain plots of input and output impedance for the network or circuit that is being analysed. Such plots, illustrated later, nearly always exhibit large changes in impedance, particularly at the turn-over frequencies of filters, and the ideal impedance value is usually taken as the infinite-frequency value, since the graph usually approximates to a horizontal line at the highest frequencies. Once this value has been established for input and output, these values can be used as terminating impedances, and Aciran used again for analysis. It is only in rare cases that you need to specify a 90° phase component for impedance, but Aciran permits this to be done, and this can be useful if the load of a circuit or its source is a capacitor or inductor.

A simple high-pass filter

The simplest form of high-pass filter is the so-called constant-k type, for which a half-section is illustrated in Figure 3.13(a). When half-sections are combined, their L and C portions will be added either in parallel or in series, but the fundamental values of L and C are taken for the purposes of calculation of terminating resistance as the values for a half-section. Half-sections can be combined either as a T-section (b) or as a Pi-section (c).

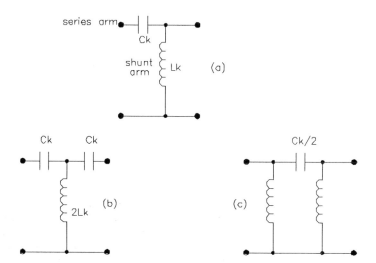

Figure 3.13 *LC filter sections. A simple half-section (a) can be used to construct a T (b) or a Pi (c) filter.*

These names are old established ones in filter theory, and are not always explained in texts dealing with electronics circuits. The constant-k type of filter uses, in its high- or low-pass forms, a single element in each arm. For example, a constant-k high-pass filter would use a single capacitor in the series arm and a single inductor in the shunt arm. It is also possible to have a constant-k bandpass or bandstop filter, in which each arm contains a single resonant circuit of opposite type. For example, a constant-k bandpass filter would contain a series resonant circuit in the series arm and a parallel resonant circuit in the shunt arm.

Theory shows that the impedances looking into each end of the half section can be calculated in terms of the frequency of a signal and a resonant frequency. If we call these two impedance values Z_1 and Z_2, then $Z_1 \times Z_2 = k^2$, so that the figure k is a constant no matter how much the values of Z_1 and Z_2 alter. As it happens, the value of k is also the value of R, the ideal terminating resistance. Note that though Z_1 and Z_2 are both impedances, k has the value of a resistance, as you would expect.

The m-derived filter uses more elaborate series or shunt sections, so that a series m-derived low-pass filter would use a single inductor in the series arm and a series resonant circuit in the shunt arm. A shunt m-derived low-pass filter would use a single capacitor in the shunt arm and a parallel resonant circuit in the series arm. The quantity m is a fraction (often taken as 0.6 in texts) which is used in calculation of component values, using as the standard values the values for a simple constant-k filter. For example, if a simple constant-k low-pass filter uses values C and L, then a series m-derived low-pass filter will use values of:

$$L_1 = mL$$

$$L_2 = (1 - m^2)\frac{L}{M}$$

$$C_2 = mC$$

$$m = \sqrt{1 - \omega_c^2/\omega_\infty^2}$$

These are for the two inductors and single capacitor in the series m-derived filter half-section. The angular frequency ω_c is for cut off, and ω_∞ the angular frequency of peak attenuation.

Terminating resistance in this respect is really a mythical quantity. The theory of a filter network of the constant-k or m-derived type is that it will have an input impedance and an output impedance. The input impedance is calculated under the assumption that the output is terminated by an assumed ideal output impedance, and the output impedance is calculated assuming that the input is terminated by the calculated input impedance. These terminating impedances are assumed to change with frequency in accordance with the theory. In practice, filters are terminated either with resistive components of constant value, or with impedances that do not accord with the theory.

For a filter of this type, the theoretical terminating resistor (valid only at the highest frequencies) would be calculated from:

$$\sqrt{\frac{L_k}{C_k}}$$

where L_k and C_k are the quantities used in the half section. In practice, filters of this type are seldom satisfactory in single sections and are usually designed as multiple sections, with half-sections used as the terminations. These calculated terminating impedances are valid only for part of the frequency range – the high frequency range for a high-pass filter and the low frequency range for a low-pass filter.

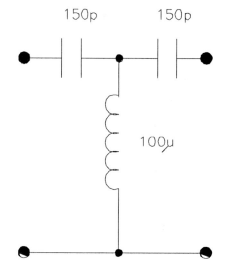

Figure 3.14 *A simple T-filter unit using 150 pF capacitors and a 100 μH inductor. The resistance and stray capacitance of the inductor have been neglected.*

Figure 3.14 shows a very simple T filter designed to work at 600 ohms nominal. This contains two capacitors in the series arms and an inductor in the shunt arm and the resistance of the inductor is usually ignored in filter calculations. We can use Aciran to see how well this circuit behaves as a filter, and also to study the effects of adding the resistance of the inductor to the calculations, and of altering the terminating resistance values.

The circuit description is fed into Aciran in the usual way for this four-node circuit, but before analysing, the *Config* menu must be used to change the *Generator* and *Load* resistor values to 600 ohms resistive. There is an option (*Impedance*) in the *Config* menu that allows the actual input and output impedance values of the network to be calculated for each frequency, but this can be ignored for the present. Once the generator and load impedance levels have been set, the analysis can be done in the usual way.

The tables shows that a good range to display is 500 kHz to 6 MHz, and the graph for amplitude response is shown in Figure 3.15. This shows a small resonant peak, as you would expect, which has been damped by the source and terminating resistor values, and the phase graph, Figure 3.16 shows the equally expected violent phase reversal which is typical of resonance. The use of the 600 ohm terminations damps the resonance enough to smooth out the response to the form of

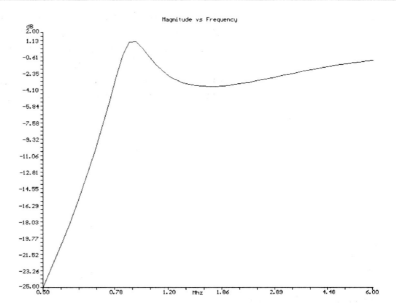

Figure 3.15 *The amplitude response of the T-filter between 600 ohm terminations.*

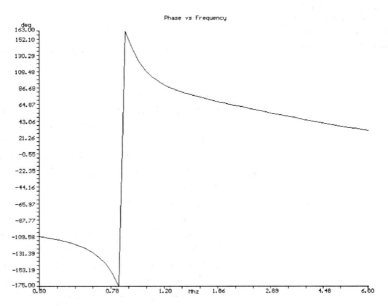

Figure 3.16 *The phase graph shows the violent phase reversal which is typical of a resonant circuit.*

the graph line that is shown, and if the default infinite generator and load resistor values are used the amplitude response (Figure 3.17) is what would be expected of an undamped circuit, with a much steeper resonance peak. The addition of any realistic value of series resistance to the inductor (winding resistance) makes no measurable difference to the graphs, as might be expected since any reasonable resistance value will have a negligible damping effect compared to that of the terminating resistors.

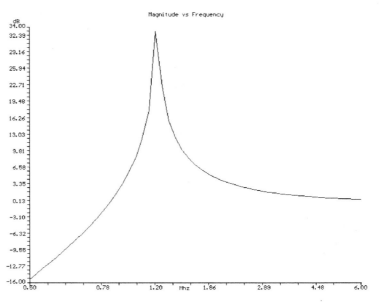

Figure 3.17 *The circuit analysed between infinite terminating resistor values shows a much more peaked response.*

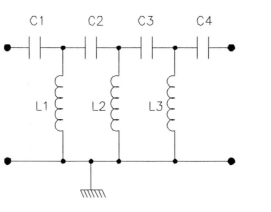

Figure 3.18 *A more elaborate ladder LC filter with three stages.*

C1,C4............560 pF
C2,C3............220 pF
L1,L2,L3.........250 μH

A constant-k high-pass filter of more elaborate design is illustrated in Figure 3.18. This uses three stages and has a ratio of inductance to capacitance that is once again aimed for working at 600 ohm impedances. Though this is a type of circuit whose analysis is available in reference books, the use of Aciran is much more revealing, with the usual ability to check the effect of changes. The first analysis uses 600 ohm impedances on each side of the filter and produces the pattern for amplitude/frequency graph as shown in Figure 3.19, using the range of 100 kHz to 10 MHz with 50 frequency samples.

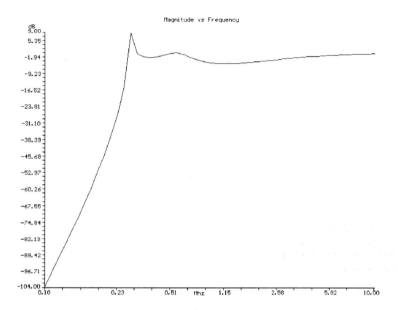

Figure 3.19 *The high-pass filter circuit analysed by Aciran, showing some nasty variations in response near the band edge.*

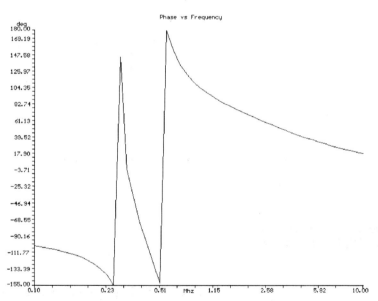

Figure 3.20 *The phase graph for the high-pass filter, showing two violent phase changes.*

This is a steep-cut filter as compared to earlier examples, but with a peak of several dB around the turnover frequency. The response near the band edge is, in fact, rather poor, with considerable variations, and the phase graph (Figure 3.20) displays violent changes of phase around two frequencies, one at 510 kHz and the other at around 280 kHz. A look at the linear plot for amplitude, using the range 200 kHz to 550 kHz shows (Figure 3.21) just how large the peak is, so that the response of this filter would be unacceptable for many purposes. This is quite

82 OTHER PASSIVE FILTER CIRCUITS

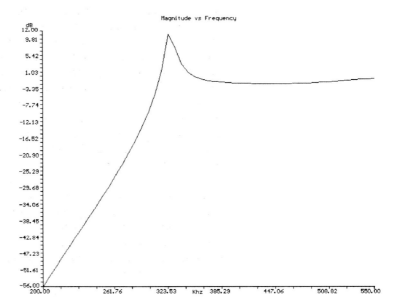

Figure 3.21 *A linear plot used to examine the band edge in more detail. This also reveals how large the peak amplitude is.*

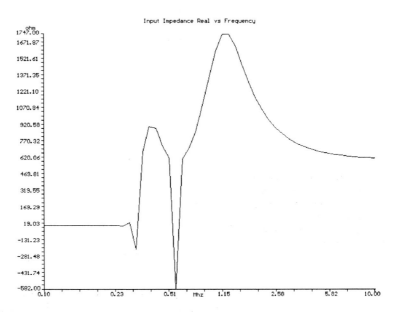

Figure 3.22 *A graph of input impedance over the frequency range, showing the actual value of input impedance is far from constant.*

commonly true of LC filters of the simpler types, and these limitations, seldom mentioned in texts, need to be realised if you want to use filters for applications in which the response near the cut-off point might be critical.

The advantage of using Aciran is that we can check some of the other factors in this filter. Figure 3.22 is a plot of the resistive part of the input impedance of the filter, showing the violent fluctuations of impedance which are typical of this type of circuit. The matching impedance that is normally used is the 'infinite-frequency' one, the value which the graph indicates will be achieved at the highest frequency ranges. This is read from the graph as about 600 ohms. The output impedance shows a similar pattern, and altering the matching impedances has little effect on the peak, which is the undesirable portion of this filter.

Using damping resistors, typically 1k in parallel with the capacitors or rather higher values in parallel with the inductors, however, does have a gratifying effect on the response, Figure 3.23. The worst of the peak has been smoothed out at the expense of a more sloping characteristic in the higher frequency range above 500 kHz, but there is still the desirably steep cut below about 300 kHz. The use of damping resistors is not covered in the formulae that are printed for this type of filter, nor is the sharply peaked nature of the undamped circuit often mentioned. Aciran allows you to experiment with damping so as to obtain an acceptable response. Note, however, that the use of damping does not smooth out the phase response to any great extent, and could not be expected to do so.

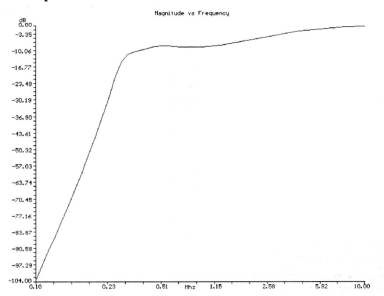

Figure 3.23 *The circuit modified with damping resistors so as to smooth down the peak response, at the expense of leaving a slow-rising response in the higher frequency ranges.*

Finally in this set, Figure 3.24 shows a very elaborate low-pass ladder network taken from a communications circuit. All of the series arms use inductors, and one shunt arm uses a single capacitor, but the remaining shunt arms use series resonant circuits. This filter is intended to work between 600 ohm terminations and the response is as complicated as the circuit diagram suggests. An analysis running over the range 10 kHz to 250 kHz shows a good low-pass characteristic (Figure 3.25) with the peaks reasonably subdued, though demonstrating the usual band-edge effects, and a steep cut beyond about 160 kHz. The extent of

84 OTHER PASSIVE FILTER CIRCUITS

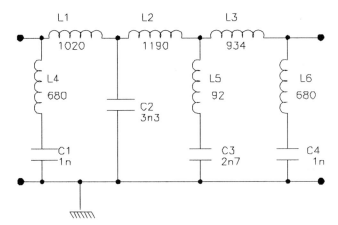

Figure 3.24 *An elaborate low-pass filter of a type used in communications circuitry, working between 600 ohm terminations.*

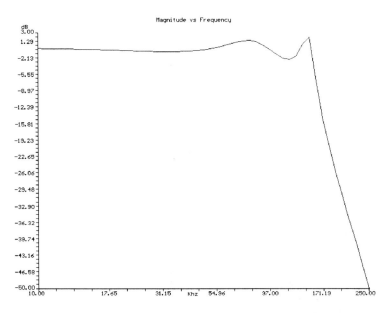

Figure 3.25 *The amplitude response of the filter obtained from Aciran.*

the cut at 250 kHz is 50 dB and the phase graph shows the usual violent phase changes that are typical of such a circuit (Figure 3.26).

Before leaving this circuit, however, take a look at a plot performed over a wider frequency range of 10 kHz to 1 MHz, Figure 3.27. This shows the characteristic that we have seen, and also an unpleasant twist in the characteristic at 366 kHz. All of this occurs at some 75 dB below the pass level, so that it is not likely to be of consequence, but it

A SIMPLE HIGH-PASS FILTER   85

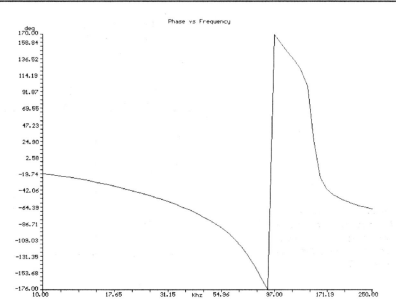

Figure 3.26 *The phase response for the telecommunications filter, showing the usual violent reversals.*

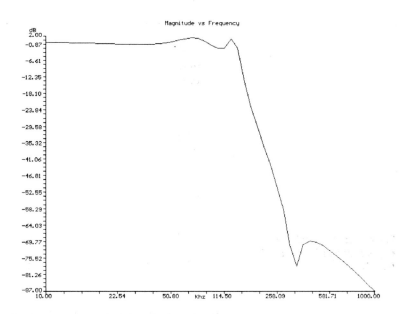

Figure 3.27 *The circuit analysis over a wider frequency sweep shows some unexpected irregularities.*

illustrates how useful it can be to investigate the performance of a circuit like this over a wide range before concentrating on the regions that are of greatest importance. The phase graph, not illustrated here, shows the expected violent phase change at the same frequency along with corresponding changes in the time delay.

Bandpass and bandstop filters

Bandpass and bandstop filters present the most formidable difficulties in analysis, and are of all filter types the most likely to produce problems in terms of unexpected responses. Taking the bandpass type first, the classic type of LC filter consists of a half-section which contains a series LC circuit in the series arm and a parallel one in the parallel arm, and this basic circuit can be used in either a T or a Pi configuration, as Figure 3.28 shows. The values of the series inductor and the parallel capacitor determine the bandwidth for a filter of this type. Both constant-k and m-derived half-sections can be designed from the data sheets that are available for filter design.

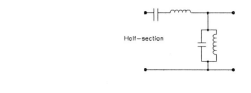

Figure 3.28 *The classic type of LC filter section and the T and Pi sections that can be derived from it.*

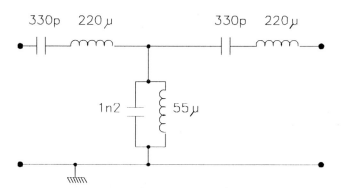

Figure 3.29 *A T-section of bandpass filter designed to work between 600 ohm impedances.*

A practical constant-k bandpass circuit is illustrated in Figure 3.29. This is a T-section with two series arms containing series resonant circuits and one parallel arm consisting of a parallel-resonant circuit, designed to work between 600 ohm impedances, and its values can quickly be copied into Aciran to produce the graph of Figure 3.30. This

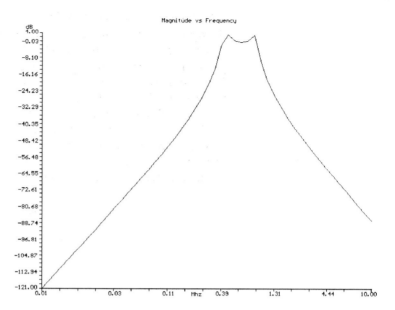

Figure 3.30 *A wide-range preliminary analysis for the bandpass circuit.*

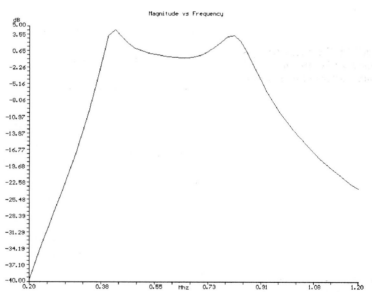

Figure 3.31 *A linear plot to emphasise the bandpass action of the circuit.*

has been deliberately plotted on a large range to see the complete picture, and it shows that a better picture will be obtained by using a limited linear range of 200 kHz to 1.2 MHz (Figure 3.31).

This plot shows that the bandpass characteristic is being obtained, but at the cost of two peaks, which for many applications will be negligible. Note also, however, that the slope of the sides is not equal; it is much steeper on the lower-frequency side than on the higher-frequency side.

The peaks can be dealt with in the usual way by adding a resistor; in this case the easiest method (not requiring any new nodes to be created) is to add a resistor in parallel with the parallel-resonant circuit, and with a 1k5 resistor the response is modified to the shape illustrated in Figure 3.32. The peak shape is still present, but smoothed to an extent that would make this filter characteristic acceptable for all but the most exacting purposes.

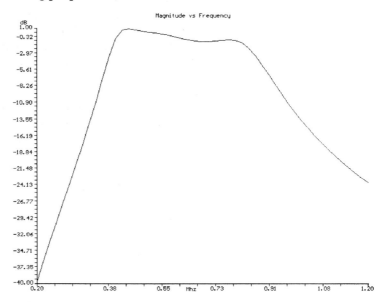

Figure 3.32 *Using damping to smooth out the peaks, in this illustration with 1k5 resistors in each parallel-resonant circuit. Note that the slopes are still unequal on this type of plot.*

The modification of a filter characteristic by adding damping resistance is something that would require considerable effort by any other method, whether from conventional analysis or by practical trial and measurement of characteristics. Aciran is particularly helpful in this respect, particularly if the damping resistance is added as a parallel resistance, as in this and previous examples, rather than as a series resistance. Using parallel resistance is not only simpler from the point of view that it uses existing nodes, it also is simpler in practical terms if the filter has already been constructed and it allows more practical ranges of values.

Bandstop filters are so similar in pattern to the bandpass type that there is no point in treating them separately. In general, the bandstop filter will use a parallel-resonating circuit in its series arms and a series-resonating circuit in its parallel arms, the opposite of the bandpass type. The older style of LC filter is being used to a lesser extent nowadays, partly because active filters using OPamps (see Chapter 5) are so much easier to design and construct, and make no use of inductors, and partly because so many circuits use digital techniques and hence use digital filters. Since digital filters are not linear circuits, their use is not something that Aciran can analyse.

More complex RC filters

The simple RC filter has a response which is not of much use because of its low rate of rise/fall from the turnover frequency, and using a circuit created from several RC stages, though this can increase the slope, causes considerable attenuation at frequencies well separated from the intended cut or boost target frequencies. There are, however, several well-known circuits using resistor and capacitor combinations which offer considerably better performance. These are circuits such as the Parallel T and Wein Bridge which are often used in RC oscillators and which are sometimes rather neglected at the higher frequencies though both circuits are used to a considerable extent for the lower frequency ranges. Both have the advantage of not using inductors, a point of considerable importance now that inductors are a threatened species. All simple passive RC circuits, however, suffer from low Q which makes their performance poor in comparison to LC circuits. The more elaborate circuits can perform rather better.

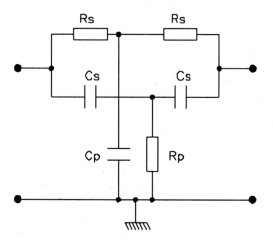

Figure 3.33 *The Twin-T filter circuit with its design parameters; this is one of the easiest bandstop circuits to design.*

$C_p = 2C_s$ and $R_s = 2R_p$

$$f_o = \frac{1}{2\pi C_s R_s}$$

Figure 3.33 shows a Twin-T circuit, with its design parameters shown. This is a bandstop type of filter which is remarkably easy to design and which can be used for frequencies in the lower RF region – stray capacitances limit its applicability to higher frequencies. The performance is illustrated in Figure 3.34 for parameters R_s = 2k2 and C_s = 100 pF, and the notable feature is how much better the performance is compared to a simple RC circuit. The Twin (or Parallel) T is used extensively as a notch filter, particularly at the lower frequencies, and its phase response, Figure 3.35, is remarkably like that of an LC circuit.

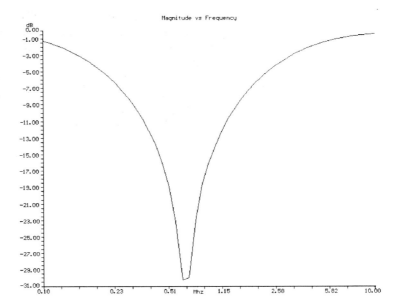

Figure 3.34 *The Twin-T analysed, an amplitude plot for values of 2k2 and 100 pF, which shows the bandstop characteristic.*

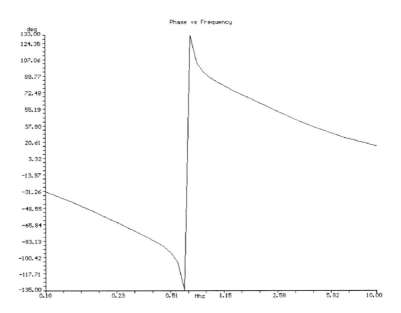

Figure 3.35 *The Twin-T phase response, showing the sharp reversal that is normally typical of resonant circuits.*

The graphs illustrated here have been taken for an analysis run using high impedance connected to each side of the circuit, and the use of a low impedance noticeably skews the response. Figure 3.36 shows the result of using this circuit between 600 ohm impedances. This point is not one that is obvious from the analysis, which ignores terminating impedances, and is one that can cause problems if this type of circuit is used between low impedances. The driving impedance is not such a problem as the load impedance, and a common solution is to drive the circuit from the emitter of an emitter follower into the base of another emitter-follower.

The Wein Bridge circuit is another bandstop type whose performance, when used in a simple circuit, is rather poor in comparison to the Twin-T type. The basic circuit is shown in Figure 3.37, with a graph for values of C=100 pF and R=2k2 shown in Figure 3.38. As the graph

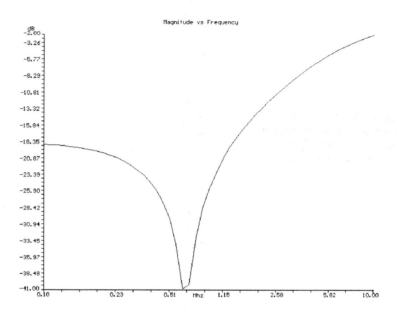

Figure 3.36 *The Twin-T used between 600 ohm impedances; the shape of the amplitude graph becomes noticeably asymmetrical.*

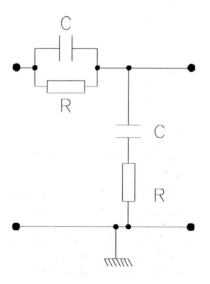

Figure 3.37 *The Wein Bridge circuit also offers bandstop action, but it needs to be made part of a bridge circuit for effective use.*

92 OTHER PASSIVE FILTER CIRCUITS

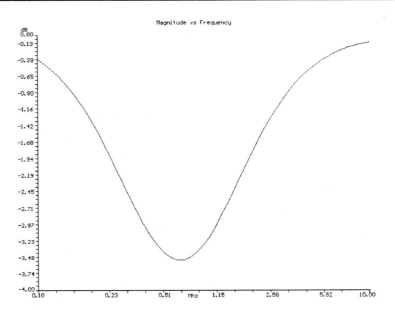

Figure 3.38 *The Wein Bridge response when used as shown. The shape is good, but the amount of cut is small. This is because the circuit is intended to be used as part of a bridge: see later.*

shows, the depth of cut when this is used as a single-ended circuit is very much less than that of the Twin-T, but what is not shown is that working into a lower impedance changes the response completely from a bandstop into a simple high-pass form of response.

The Wein Bridge is, however, not intended to be used in this form and though a small amount of bandstop action can be obtained, the circuit is really intended to be used in a bridge circuit, balanced against two resistors with a 2:1 ratio. This form of use gives very much better results with a steep-sided response and a much larger attenuation at the cut frequency. This form of the circuit can be simulated only by using a phase splitter, and Aciran, as shown in the following section, can simulate a perfect transformer to provide the phase-splitting action.

Using transformers

Aciran provides a transformer as a component, allowing a number of circuit dodges to be used such as the phase-splitting action noted above. The Aciran transformer is a theoretically perfect one whose turns ratio determines its voltage ratio, and which has no losses of any type, no inductance and no capacitance. For many of the practical uses of transformers the losses must be simulated by using inductance and resistance values, but for some purposes the perfect component is useful to simulate effects that otherwise would be difficult.

Note that Aciran can simulate a fully-floating transformer with no connections to earth. The instruction documents for Aciran refer to the older version, which insisted that one terminal of both primary and

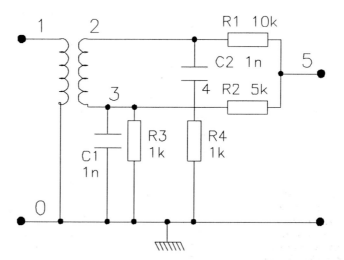

Figure 3.39 *A Wein Bridge circuit fed from a transformer to illustrate the use of a transformer in Aciran and also to show the Wein action.*

secondary must be connected to Node 0. This is no longer necessary, and it allows the transformer to be connected so as to supply a floating-signal source to circuits such as the Wein Bridge (Figure 3.39).

The circuit in this example shows the node numbers to illustrate that a transformer is configured with four node numbers. These node numbers, along with the ratio of primary to secondary turns, define the perfect transformer completely, and in this example the ratio is 1:1. The essence of the Wein Bridge used in this way, is that the resistor connected to the series arm must be of double the value of the resistor connected to the parallel arm.

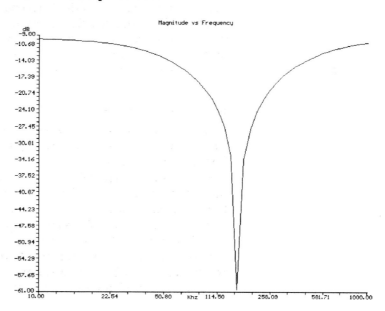

Figure 3.40 *The response of the Wein Bridge circuit, showing the impressive amount of cut and the steep-sided response.*

The performance difference between this and the earlier example is quite impressive, as Figure 3.40 shows. The amount of cut at the mid-band frequency is now around 60 dB, and the steepness of the cut is impressive by any standards for RC or LC circuits. The frequency of maximum cut is calculated from the equation:

$$\frac{1}{2\pi RC}$$

This assumes that the values of R and C in the bridge are equal, and that the resistors R_1 and R_2 are in the ratio 2:1.

This illustrates the use of the transformer component in an ideal way – no-one would normally construct a Wein Bridge circuit using a transformer as a split-phase supply, though the principle is quite feasible and could be used at the higher frequencies. The point here is that the ideal transformer can supply an action – that of providing a fully-floating pair of phase-inverse inputs – that is not otherwise available without using active components. Wherever such an input is needed, it can be simulated in this way.

The more normal action of a transformer, however, is to provide the impedance matching or voltage/current level changes in a circuit. In some cases, treating the transformer as a perfect component is quite reasonable but it is more likely that you will need to insert the additions that allow for the inevitable imperfections in order to analyse the circuit in a practical way. This point is dealt with in Chapter 6.

A centre-tapped transformer, common in audio circuits at one time and still used in communications circuits, can be simulated by using two transformers connected together. Figure 3.41 shows the equivalent

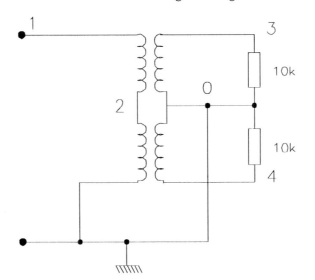

Figure 3.41 *Using two interconnected simple transformers to simulate the effect of a centre-tapped secondary.*

circuit which can be used by Aciran, with resistive loads that are used in the analysis. This analysis must be done twice, once for each output, altering the number for the output node in each example. The phase difference of 180° will be shown; the output at Node 3 is 0° and the output at Node 4 is 180°.

☐ Note that you will get problems unless you use realistic generator and load impedances – the program will analyse but will refuse to graph and it ends with an error message. In this example, a generator impedance of 10k and a load of 100k was used.

Since the transformers used in this example are supposed to be perfect, the graph shows no change with frequency, and only the losses due to the resistors are shown.

4 Using active circuits

Transistors and FETs

Active circuits are circuits which contain components that produce power gain, such as transistors (bipolar or FET) and operational amplifiers. Aciran can deal with all of these components, but not with specialised integrated circuits whose characteristics cannot be described by the formulae that are used for the listed devices. The position of an active device in a circuit is dealt with by Aciran in much the same way as any passive component, so that you will need to supply the node connections for the three terminals. For the bipolar transistor these are base, emitter and collector; for the FET they are gate, source and drain. The OPamp requires nodes for its positive input, negative input and its output only, since biassing or power supply connections play no part in the signal path.

☐ Specialised ICs can be dealt with using Aciran's current or voltage generator equivalents, but only if you can obtain a formula or equivalent circuit for the overall action.

Active circuits are analysed by Aciran on the basis of tables of active components, or by filling in values on a form. The component tables are held as files, and by selecting a file, all of the data on an active device can be read by Aciran. You have to select the file each time a given device is used; there is no provision for notifying to Aciran that all active devices are to be of the same type. The option of filling in values is slower, and if the same device is to be used several times over the same data will have to be typed several times over. This option is not always very useful for bipolar transistors in particular because the information that is needed is often quite difficult to obtain, though FETs are easier. Data on operational amplifiers is fairly easy to obtain, but is not necessarily the data that Aciran requires.

Transistors

As it comes to you, Aciran contains data for 11 bipolar transistor types and 11 FETs. The transistor is modelled in a fairly simple way, using three nodes and allowing for parameters of h_{ie}, h_{oe} and h_{fe}, along with the tolerance of h_{fe} and the turnover frequency f_T. A Hybrid-Pi model can be used, and is described in the Aciran manual, but its refinement is useful only if you are using the transistor near its frequency limits, and you have access to all of the data that is required. For many purposes, the transistor plays no part in the frequency response of the circuit and the simple model is perfectly adequate. When a transistor is specified you must pick from the set that is available, or designate characteristics for yourself.

- ☐ It is not easy nowadays to find tables of transistor or FET characteristics of the sort required for Aciran. The tables of transistor characteristics that are so readily available for bipolar transistors carry only the h_{fe} values, but not the h_{ie} and h_{oe} values. This encourages you to analyse circuits using the models supplied, or to work with operational amplifier circuits as far as possible.

The set of transistor characteristics includes one labelled as Standard which can be used as any general-purpose silicon transistor. Because the model is based on the well-known G_m equation (G_m = 40 mA/V per mA of DC collector current) you will be asked for the collector current I_c as well as for the connection nodes for the transistor. Since the Aciran examples include a straightforward single-stage amplifier there is no point in repeating this example here, and we can concentrate on circuits which in practice cause more trouble. In the examples that follow, the models that are supplied with Aciran will be used (even where they are not strictly applicable) in order to indicate the use of these models.

When *Transistor* is selected from the list of components, you will be asked by a message at the foot of the screen if you want to use a model. The default answer is Y and if you press the ENTER key, you will see either the list of models (if they are in the same directory) or a list of directories, from which you can select the Models directory so that the names of the models appear. The required model is selected in the usual way by placing the selection bar on it and pressing ENTER or by using the mouse button.

- ☐ If you do not opt for a model that is already stored, you will be asked to supply parameters for the transistor you intend to use. These parameters are not necessarily easy to find, and they have to be entered into a form (see later). You should add to the existing models list any transistors you use regularly.

Circuits for analysis can be rather simpler than might be expected, because bias components need not be shown unless they are in the signal path or a feedback path. As far as single transistor circuits are concerned, this makes no difference, but for circuits such as the long-tailed pair, the bias for the base of the output transistor Tr2 (Figure 4.1), can be ignored providing you can specify what the DC collector current will be, and this connection can be taken to Node 0. In this example, stray capacitances have been represented as components C_s to make the simulation more realistic and no connection is indicated for the base of Tr2, since this will be a steady voltage.

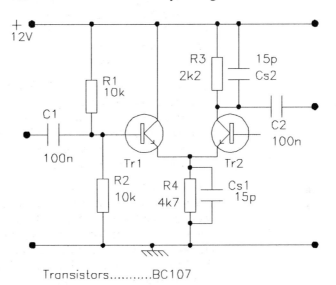

Figure 4.1 *A long-tailed pair circuit in which the connection for the base of Tr2 can be given as Node 0. Note the stray capacitance values.*

☐ You nevertheless need to know the steady collector current for each transistor in the circuit, because this affects the signal gain. For a circuit such as the long-tailed pair, the I_c value must be a realistic one based on the values of resistors in the collector and emitter circuits of the transistor.

The graph for 50 Hz to 10 kHz, Figure 4.2, shows the poor low frequency response due to the 100 nF coupling capacitor along with the 10k bias resistors at the input. The current in each transistor has been specified as 0.574 mA, determined by the size of the common emitter resistor R3. The mid-band gain is of the order of 37 dB, and trial runs at higher frequency ranges show that this is held to around 760 kHz, normal for this type of circuit with the sizes of stray capacitances indicated. A similar value of stray capacitance at the input will reduce the high-frequency 3 dB point to some 700 kHz.

A circuit which is often difficult to deal with is the common combination of two transistors into a cascode circuit to feed an inductive or resonant load; it can be used for any high-impedance load. The difficulty as far as Aciran analysis is concerned is in the specification of collector

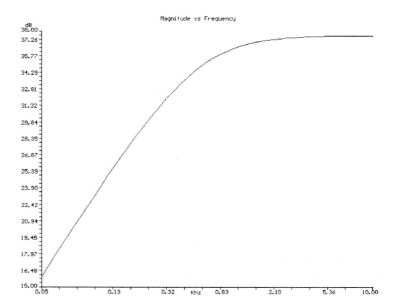

Figure 4.2 *The amplitude frequency graph for the long-tailed pair circuit, showing a poor low frequency response in this particular example.*

current, because this depends critically on the bias components and the h$_{fe}$ values for the first transistor. Figure 4.3 shows a typical cascode circuit for driving an inductor of fairly large value. Since the load is an inductor with stray capacitance forming a resonant circuit, the gain should show a peak at the frequency of resonance and a quick calculation shows that this should be around 34 kHz.

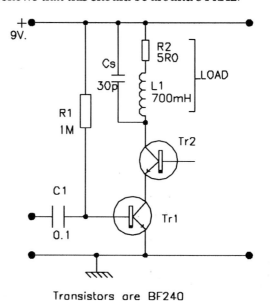

Figure 4.3 *A cascode amplifier circuit with an inductive load. Once again, the base connection of Tr$_2$ is not indicated since it is fixed and can be taken as Node 0.*

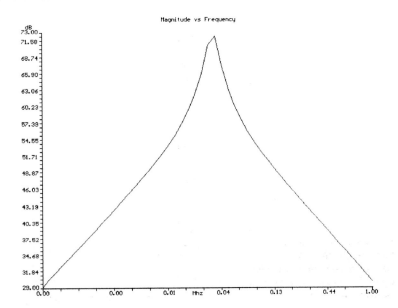

Figure 4.4 *The amplitude frequency graph for the cascode circuit, which shows the prominent effect of resonance.*

Figure 4.4 shows the analysis for the frequency range of 1 kHz to 1 MHz. The resonant peak is very pronounced in this plot, and it is tempting to see what might be done to reduce this with a damping resistor. Using a 10k damping resistor connected across the load alters the characteristic to that illustrated in Figure 4.5, with a maximum gain of around 36.3 dB. The bandwidth cannot be measured on this logarith-

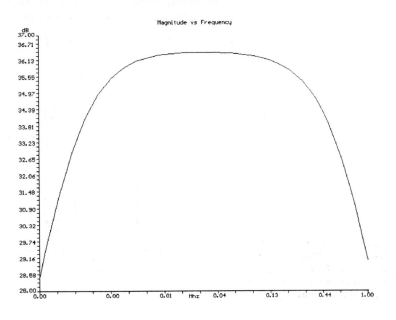

Figure 4.5 *Using a damping resistor greatly flattens the response curve at the expense of gain.*

mic plot (the lower values of frequency are all shown as zero) but on a pair of linear plots – one for low frequencies, one for high frequencies – the 3 dB bandwidth can be seen to run from 2 kHz to 500 kHz. Whether the circuit is used with this extra damping or not depends on the requirements, and you can make use of the trade-off between gain and bandwidth as you please. Remember, however, that the performance of the circuit depends heavily on the value of its bias current, so that the value of R_1 and supply voltage, along with the actual h_{fe} value for the transistors (usually around 60 for the BF240) will affect the gain considerably.

This circuit also illustrates another use for Aciran in calculating input and output impedances. By checking the *Impedance* line in the *Config* menu, you can obtain graphs of the input and output impedances for the circuit in the presence of the selected loads. If the driver and load impedances are maintained at the default 100M values, the impedances that are measured in this way are the true input impedance and the output impedance with (in this example) the inductive load. The graphs show that the input impedance is steady at around 8k5 for most of the range of frequencies, with the output impedance showing the typical resonance curve. If a resistive load is used, the output impedance is virtually that of the load resistor, since the circuit itself has a high output impedance.

The use of Aciran with transistor models can be very helpful in analysing negative feedback circuits, and the two familiar two-transistor types form a useful illustration of the analysis for gain and impedances. In the following illustrations, conditions have been arranged in many of the examples to permit the use of 1 mA operating currents, though this is not necessarily optimum, particularly for the first transistor in each pair. Figure 4.6 is the circuit for a series-derived shunt-fed pair, using typical values. As usual, we can simulate the effects of stray capacitance (not shown in the diagram) by connecting 10 pF capacitors from each input and output node to earth, and this has been done in the Aciran simulation. The bias components R_2, C_2 are included because they will affect the low-frequency response, but if this were of no concern, the circuit could show the emitters of Tr_1 connected directly to earth for AC signals.

The gain/frequency graph, Figure 4.7, shows the usual response over a fairly wide range, with a maximum gain of 63 dB, though the low-frequency 3 dB point is unusually high at around 5 kHz because of the very low input impedance of the first transistor, around 15 ohms, and the limited decoupling for the emitter resistor. The higher 3 dB point is determined by the strays, assumed to be of the order of 10 pF and is around 220 kHz. The input impedance is around 15 ohms for most of the working frequency range, rising at the extremes, and the output impedance is constant at around 3.9k for all frequencies up to the high limit. Since the actual resistor at the output is 3k9, this indicates that the intrinsic output impedance is considerably higher, as theory predicts.

TRANSISTORS 103

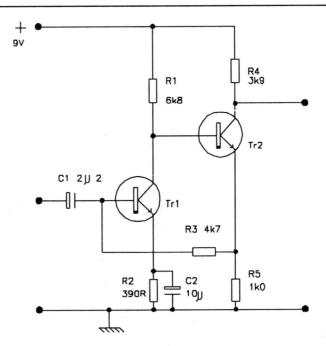

Figure 4.6 *A negative feedback pair, using the series-derived and shunt-fed type of circuit.*

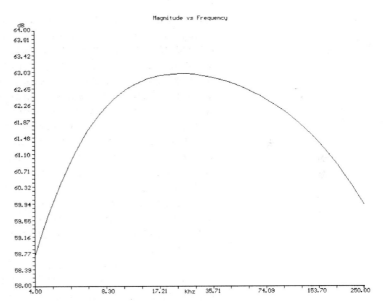

Figure 4.7 *The analysis of the feedback pair shows a far from ideal shape.*

The low-frequency response would be improved by using a large value of coupling capacitor and a lower current in the first transistor.

The other common two-transistor feedback arrangement is the shunt-derived series-fed type as in Figure 4.8. The feedback resistor is R5, 47k, so that the shunting of the load resistor R6 is not significant. The

104 USING ACTIVE CIRCUITS

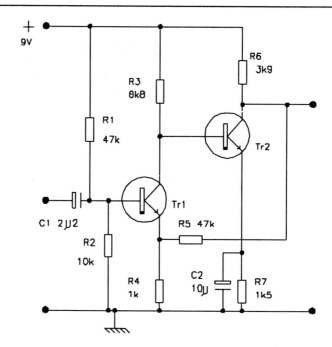

Figure 4.8 *Another form of a feedback pair using a shunt-derived series-fed circuit.*

resistors R_1 and R_2 establish the base voltage for Tr_1, and must be shown in the simulation because they shunt the signal. The bias components for Tr_2, C_2 and R_7, are included because of their effect on the low-frequency response, but could have been omitted in this example.

The analysis of this circuit for frequency response, Figure 4.9, shows a much-improved low-frequency characteristic down to almost 10 Hz,

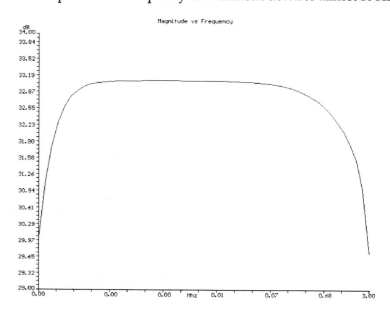

Figure 4.9 *A much better response shape from the shunt-derived series-fed type of circuit.*

with the high-frequency 3 dB limit extending to almost 3 MHz. One of the attractive features of this type of circuit is that though it needs rather more components than the previous version, its impedances are better suited to most applications and it is easy to get a good wide-band response. The mid-band input impedance is 8.2k and the mid-band output impedance is 3k, only slightly less than the load presented by R_6 and R_5.

Straightforward feedback amplifier circuits are not the most pressing applications for Aciran, however, and circuits that incorporate reactive components in the feedback path are of much more interest. Such circuits are used as methods of matching a required response curve or for compensating for loss of gain at high frequencies. The more elaborate the feedback network, the more useful Aciran is in determining the overall effect.

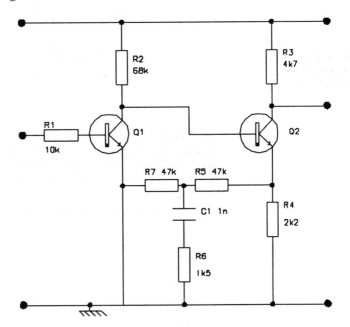

Figure 4.10 *A two-transistor amplifier whose negative feedback path includes a filter network. Bias for Q_1 not shown.*

Figure 4.10 shows a feedback amplifier which uses a filter network in the feedback path. The first transistor is being run at 0.1 mA with a high collector load and direct-coupled to the second, from whose emitter the feedback connection is taken. This type of circuit can be analysed approximately with a few calculations, but an analysis by Aciran reveals much more than can be easily found in the same time by any other methods. The STANDARD transistor model of Aciran will be used in this and the next two circuits for the sake of illustration.

The amplitude graph of Figure 4.11, plotted here as a logarithmic sweep over the range 100 Hz to 1 Mhz shows a gain of around 26 dB at low frequencies, increasing, with a turnover frequency of about 70 kHz,

106 USING ACTIVE CIRCUITS

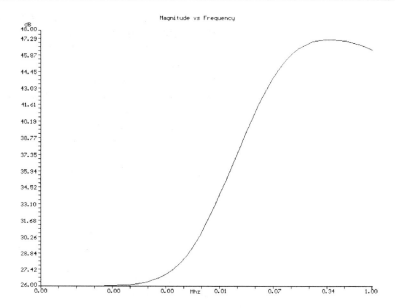

Figure 4.11 *A tailored response obtained by using the network in the feedback path.*

to around 47 dB at around 300 kHz, with a slight downturn to 46.35 dB at 1 MHz. Altering R_6 will change the value of peak gain, and altering C_1 will change the turnover frequency.

Though this graph is fairly predictable, the phase graph of Figure 4.12 is less so, and is surprisingly symmetrical. Rough calculations on feedback-circuit performance generally neglect phase shift, and using Aciran in this way can often be extremely useful, particularly if the

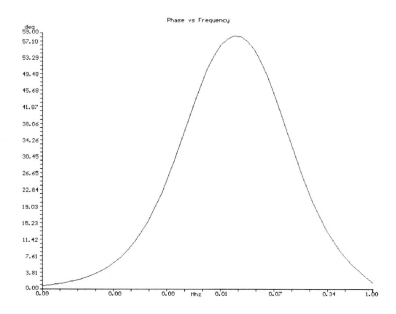

Figure 4.12 *The phase response for the same circuit, showing a peak of around 59° at 30 kHz.*

feedback is being taken from the output. By checking phase shift with the feedback unconnected, for example, it is possible to see if there is any possibility of the negative feedback reversing at any frequency (unlikely in a two-transistor circuit with no inductors) and so causing instability. Remember that Aciran allows you to choose which point to take as an output, so that you can, in this example, take the emitter of Q2 as the output, disconnect R5, and analyse the circuit for phase shift and gain. As we shall see, Aciran does not indicate the onset of instability by showing infinite gain.

An analysis of input and output impedances is often very useful in amplifiers like this, and when this choice is switched on in the *Config* menu of Aciran, with the nominal drive and load impedances set at the usual 100M figures, tables and graphs of input and output magnitude and phase angle can be obtained. When this is done for the circuit of Figure 4.10, the input impedance magnitude is shown to be fairly constant at just over 10k, with a rise in magnitude starting at 10 kHz up to 10.9k at 1 MHz. The input impedance phase is zero at low frequencies, attaining a maximum of 1.55° at about 1 MHz. The output impedance magnitude remains constant at 4k7, with a phase angle that is also fairly constant to within 0.1°.

Figure 4.13 illustrates a circuit whose feedback is taken from the output point and whose frequency response is shaped by another RC circuit. The emitter of Q2 is taken as being connected to earth, Node 0, because of the decoupling – this should be done unless you are concerned about the effect of the decoupling time constant on low-frequency response.

This circuit has two sets of time constants that will affect the high-frequency response, so that a more widely-ranging graph is needed.

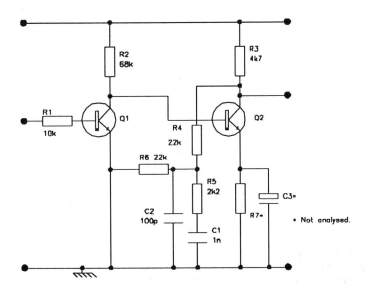

Figure 4.13 *Another form of feedback circuit with shaping components in the feedback line.*

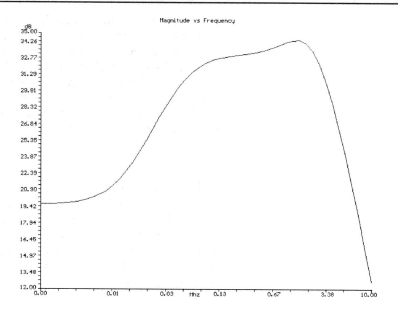

Figure 4.14 *The complex response curve for the circuit of Figure 4.13.*

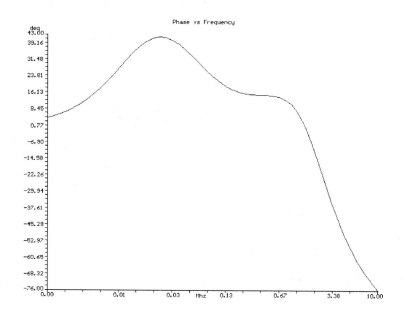

Figure 4.15 *The phase-shift plot shows how small are the changes up to several MHz.*

The result of a sweep from 1 kHz to 10 MHz is shown in Figure 4.14, showing the low-frequency gain of around 19 dB rising to 32 dB at 130 kHz, with a turnover to a smaller slope, and a final sharp turnover at about 3 MHz, with a very steep fall thereafter. The phase diagram, Figure 4.15, shows surprisingly small phase changes over the main part of the frequency range, due to the feedback being taken from the output terminal of the circuit.

There are more surprises in the input and output impedance values. The input impedance starts at more than 5M at low frequencies, falling in a graph shape that almost mirrors the gain curve to about 100k at 10 MHz. The phase shift of input impedance is very complex, with three sinusoidal cycles of variation, but all between the limits of –2° and –30°. The output impedance magnitude starts at around 1k5, rising to a limit of 2k68 at 10 MHz, and its phase angle peaks at 13.5° for the frequency of around 20 kHz, remaining below this value for all other frequencies.

Positive feedback

Aciran can deal with circuits that contain positive feedback providing that this does not cause instability. Since it is important to know what Aciran does when a circuit becomes unstable at some frequency, a look at two-transistor circuit which use some positive feedback can be useful, particularly since this also offers another method of introducing peaks into the response curve.

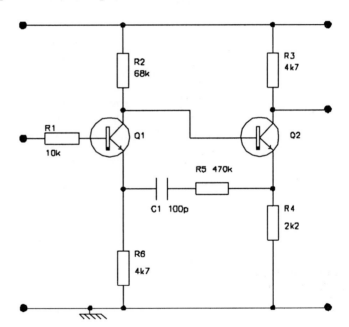

Figure 4.16 *A circuit which provides positive feedback – try this out so that you can recognise the signs of instability, since Aciran does not indicate directly when a circuit will oscillate.*

Figure 4.16 shows a circuit in which positive feedback has been introduced between the emitter loads. The amount of feedback is controlled by the ratio of R_5 to R_6, and the turnover frequency is determined by the time constant of C_1 and R_5. The effect, as Figure 4.17 shows, is to produce a wide-band response with only a 1 dB peak. The amount of positive feedback is small in this example, and it is being

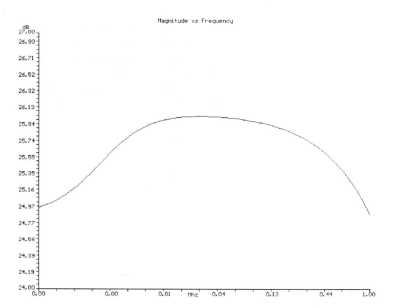

Figure 4.17 *The response of the circuit as drawn. The positive feedback contributes little to the gain, creating only a mild hump in the response.*

applied over a wide range of frequencies (the impedance of C_1 equals the resistance of R_5 at around 3.3 kHz).

If the value of R_5 is reduced to 100k, the gain graph becomes much more steeply curved, rising to 31 dB at around 150 kHz (Figure 4.18). The positive feedback is now greater in magnitude and the turnover frequency is higher because of the change in this resistance value. The

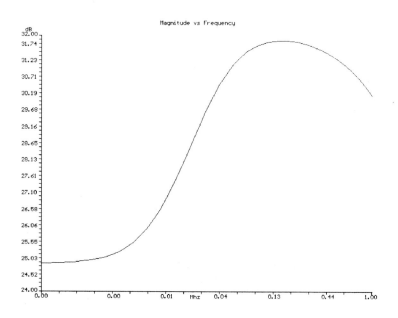

Figure 4.18 *With lower resistance, 100k, in the feedback path, the peak in the response is more prominent.*

phase graph, not illustrated, shows a peak of 23° at 30 kHz, but remains low at other frequencies.

Altering the value of R5 to 50k causes a more dramatic set of changes. The gain now peaks very sharply (Figure 4.19) at around 500 kHz and 66 db. This response would not normally be useful for any purposes, but with reduced time constants, the peaking could be useful for frequency compensation. The phase graph of Figure 4.20 displays a very severe reversal, swinging from almost +170° to −85° in a very small frequency range.

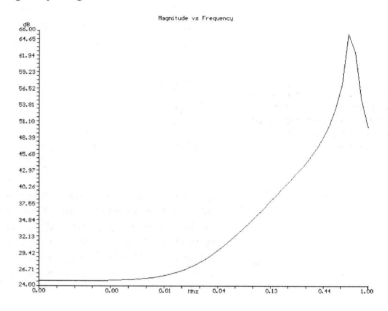

Figure 4.19 *The peak becomes very sharp with a resistance of 50k. The circuit would, in fact, be oscillating at this stage.*

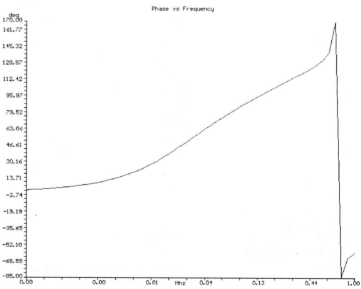

Figure 4.20 *The very severe phase reversals are another indication of instability in the circuit.*

As it happens, Aciran does not show in any dramatic way when a circuit becomes unstable by, for example, indicating infinite gain at some frequency. The graphs in Figures 4.19 and 4.20 are the tell-tale, because the violent phase reversal and very peaked magnitude graph appear at the point when the positive feedback would start to cause oscillation. In this particular circuit, the phase angle of the input impedance also shows large swings in value.

Remember that positive feedback is normally undesirable because it makes the amplifier less stable and more affected by component tolerances and other variations. As a way of compensating for the effect of strays, a limited amount of positive feedback over a limited frequency range can be useful. It is only for such circuit applications also that Aciran can be helpful, because the output of Aciran for circuits in which positive feedback is applied over a wide frequency range can be misleading, particularly when the amount of feedback is enough to cause instability.

Audio active circuits

Other types of active circuit that lend themselves well to analysis by Aciran are the audio equalising circuits and active tone controls. These are often illustrated with very idealised drawings of their responses, so that closer analysis can often be useful to determine the effect of more realistic conditions, particularly of loads.

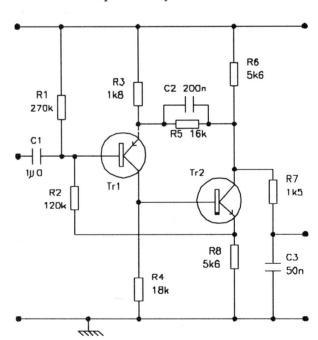

Figure 4.21 *An input stage for magnetic pickup cartridges. The bias and decoupling components which are not in the signal path have been omitted.*

Figure 4.21 shows a conventional transistor input stage for magnetic cartridges in the pre-CD days. This uses a PNP-NPN pair in a series derived shunt drive circuit which is designed with time constants to compensate for both the characteristics of the cartridge and the way in which the signal levels are doctored before recording to reduce the effects of surface noise. The output characteristic should be as shown in Figure 4.22, an idealised set of curves smoothed into a single curve.

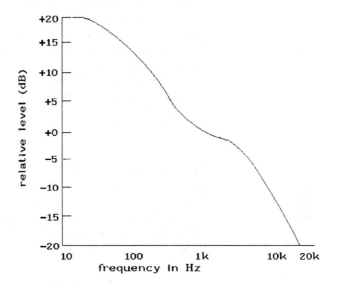

Figure 4.22 *The ideal output characteristic for a magnetic pickup cartridge (RIAA characteristic).*

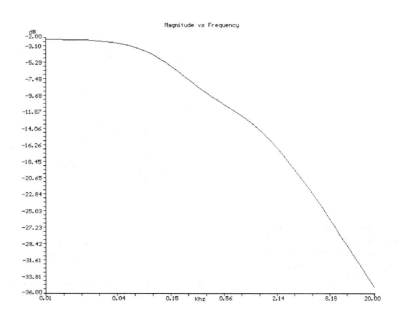

Figure 4.23 *The output characteristic as produced by Aciran.*

114 USING ACTIVE CIRCUITS

As it works out, the shape of the graph of amplitude plotted against frequency looks very convincing (Figure 4.23). The absolute decibel quantities are not as laid out in the specification, however, and the input impedance is higher than it ought to be for a magnetic cartridge. Ideally, the input impedance should be around 50k, but for this circuit it fluctuates slightly around the 240k mark over the audio range. The output impedance, dominated by the capacitor C_3, is around 5.7k at low frequencies, dropping to 158R at 20 kHz. This preamplifier would normally be connected into a circuit with high input impedance.

Just before the advent of the CD, the development of black disc recordings had led to the increasing use of moving-coil pickups with very low output and very low impedance levels. Preamplifiers were needed for these pickups, and a design due to John Linsley Hood is illustrated in Figure 4.24. This would normally be constructed using selected low-noise transistors, but is shown here using rather less exotic types drawn from the Aciran standard set of models. The capacitor C_1 is regarded as a short circuit from the signal point of view, and bias conditions have not been precisely set. The circuit would normally be operated from a low-voltage supply of the order of 3 V with very low currents. In the example, the transistors are assumed to operate with currents of 180 µA, 300 µA and 1 mA respectively. Note that Aciran requires currents in mA, so that these quantities are entered as 0.18, 0.3 and 1.0 respectively.

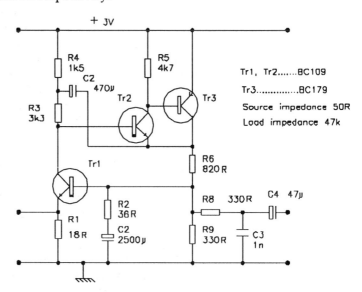

Figure 4.24 *A moving-coil pickup preamplifier design by John Linsley Hood. An Aciran analysis reveals its impressive performance.*

The results from this circuit, as we would expect of any circuit from John Linsley Hood, are very impressive, as Figure 4.25 shows. The peak gain is 50.45 dB and at the low frequency limit of 10 Hz this is down by only 0.09 dB. At the high frequency limit of 20 kHz, the response is down by only 0.349 dB. No account has been taken of stray capacitan-

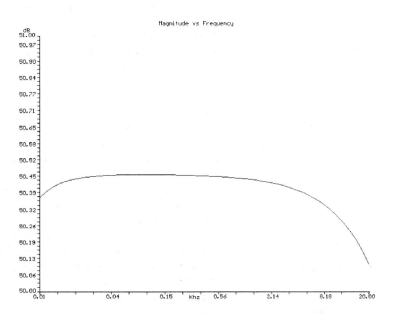

Figure 4.25 *The Aciran plot for the preamplifier, showing a remarkably flat characteristic.*

ces, but since the circuit impedances are so low there is little point in re-testing with typical values of stray capacitance.

Cassette preamp circuits are another example of complexity which is brought about by the need for equalisation. A set of signals recorded at the same level of magnetization (flux level) will cause output from the replay head whose amplitude is proportional to frequency except at the extremes of the range, so that a replay amplifier needs to have a falling response over the main part of the frequency range. This needs to be modified at both the low- and the high-frequency ends, particularly for high-frequency signals which are not efficiently recorded nor reproduced because of the finite size of the gap in the record/replay head.

A circuit formerly used in Sony machines is illustrated in Figure 4.26. This consists of a direct-coupled triplet of two PNP and one NPN transistor with a negative supply rail to maintain bias levels, and with extensive and elaborate feedback loops. The circuit is unusual in being a form of cascode in which Tr_3 is the only transistor with a connection to the positive supply line. The base current of Tr_3 is the only source of current for Tr_1 and Tr_2, and this makes it difficult to allocate sensible values of collector current for an analysis by Aciran. Circuits like this can be very difficult to analysis unless you possess a service sheet or other document which shows the transistor currents or provides voltage readings from which the currents can be calculated.

Making guesses that appear to be reasonable for collector currents, the analysis can be carried out in the usual way, remembering that the negative supply line is another Node 0. The graphical analysis of

116 USING ACTIVE CIRCUITS

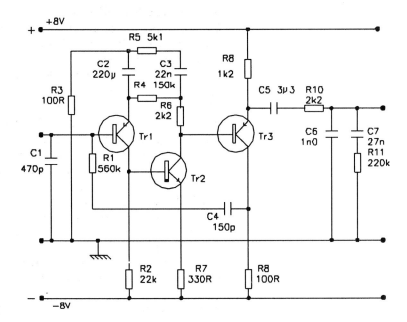

Figure 4.26 *A tapehead preamplifier design due to Sony which uses the base current of Tr3 to set the collector currents for Tr1 and Tr2.*

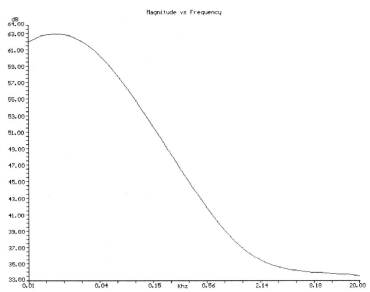

Figure 4.27 *The output characteristic for the tape preamp which follows almost exactly the theoretical requirement.*

amplitude plotted against frequency is shown in Figure 4.27. This shows an excellent response curve with maximum gain in the bass and following the recommended curve shape to a gain of some 33 dB at 20 kHz. The circuit is designed to use low-noise transistors, and by running the first two stages at such low currents the noise is reduced further. Aciran in its present version, however, cannot analyse for noise levels.

Wide-band amplifiers

The analysis of wide-band amplifiers is always difficult and the assistance of Aciran is very valuable, although much of the difficulty of predicting performance lies in determining the effect of stray capacitance and inductance. Reasonable allowances can be made for stray capacitance, but stray inductance is much less easy to account for. The advantage of using Aciran is that it becomes possible to allow for best and worst cases by assigning large tolerances to all stray values and running a tolerance analysis.

Broad-band amplifiers use either feedback with capacitive peaking, or the use of inductive peaking, or both. The results that are obtained will depend very much on what is assumed for stray capacitances, particularly for inductively-peaked circuits which will not display any resonance effect unless strays are included to resonate with the inductor(s). It is particularly important to place the strays in the correct places – the stray capacitance between collector and base is much more important, because of its Miller feedback effect, than stray capacitance between either terminal and earth.

A single video amplifier stage from a long-forgotten TV receiver is shown in Figure 4.28. This has been simplified to omit all components connected to the base, because the components consisted in the original of a large-value resistor connected to the positive supply, and a diode connected to a varying level set voltage. As far as signal voltages are concerned, these can be ignored. Diodes must be ignored in any case, or substituted by resistors for the sake of the analysis, since a diode is

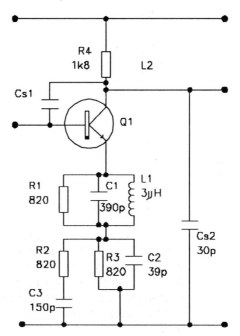

Figure 4.28 *A typical old-time colour TV video amplifier stage whose performance would be difficult to assess without Aciran.*

essentially a non-linear component. A stray capacitance of 10 pF has been assumed between collector and base, and a 30 pF stray at the load. The source resistance is set to 200R and the load resistance to 2k2 to simulate the conditions under which this circuit normally operates.

The resulting gain frequency graph, Figure 4.29, shows the steady rise in gain from 100 kHz up to the position of the 4.43 MHz notch filter (L_1C_1). The peak gain is at about 1.8 MHz and is at that frequency some 4 dB greater than the gain at the lower frequencies. This sort of non-uniformity of gain is not unusual in TV video amplifiers and as long as there are no sharp changes in gain, other than for notch filtering purposes, there is little point in pursuing the perfection of a straight-line response – it wouldn't make Wogan look any different.

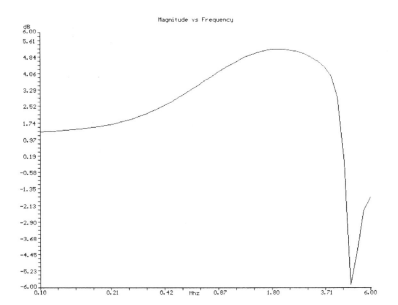

Figure 4.29 *The amplitude plot for the video amplifier shows a rising characteristic at high frequencies up to the frequencies covered by the notch filter.*

A simple wide-band type of power amplifier is shown in Figure 4.30. This is intended to operate between 50 ohm impedances, so that the *Config* menu of Aciran should be set for this level at both input and output. Both of the transistors are operated at high currents, with the input transistor at 20 mA for best matching of input impedances. The Aciran plot will show a voltage loss rather than a gain, but the current gain is enough to ensure that there is a power gain. The input stage is common-base, and it feeds an emitter follower to provide the low-output impedance. Strays of 10 pF each have been included (shown dotted) in the diagram as a reminder that their inclusion is needed to make the analysis realistic. The action of the circuit is critically dependent on the values of C_2 and R_4, of which R_4 affects bias as well as signal conditions.

WIDE-BAND AMPLIFIERS   119

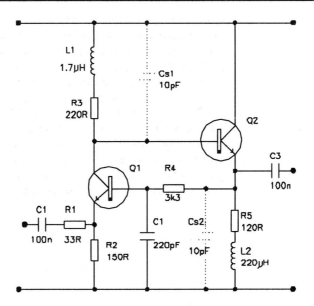

Figure 4.30 *A wide-band amplifier intended for power gain and working between 50 ohm impedances.*

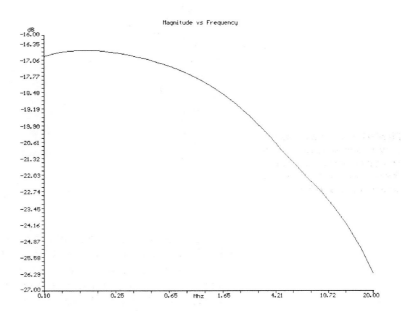

Figure 4.31 *The response of the broad-band power amplifier – remember that this is a voltage gain plot which ignores current and power gain.*

The amplitude/frequency graph is shown in Figure 4.31. Though this would not be regarded as an impressive performance for a voltage amplifier, the power output of the circuit, plotted against frequency, is considerably better than this graph would indicate. Unfortunately, Aciran cannot provide figures for power gain, and its main application to circuits of this type is to ensure that circuit values are well matched to the application. If, for example, the circuit were to be used between different impedance values it would be possible to experiment with different values of R3, R4 and C2 to try to ensure as uniform a voltage gain graph as possible.

Power amplifiers

Power amplifiers are not an obvious application for Aciran because most power amplifiers work in Class B, which is not a linear mode and because Aciran, as we have seen from the example above, does not indicate power gain. The use of Class B, however, does not result in any cut-off of the signal at the output, and since Aciran ignores bias conditions apart from transistor steady-state settings, it can be used to find overall gain and, just as important, the effect of networks such as the Zobel network on the output at the loudspeaker. It is also possible to include the cross-over network as part of the output stages and show the overall frequency and phase response.

The difficulty here is information. The set of models for Aciran does not include any power transistors, and tables do not show values of h_{ie} and h_{oe} for power transistors: see below.

Adding transistors to the set

The small set of transistors modelled by Aciran can be added to indefinitely for as long as you have disk space for files. Since a transistor description file takes only around 40 bytes there is clearly room for several hundred transistor models on even a floppy disk. The information that needs to be supplied is the same as will be asked for if you specify not to use the models. It is:

h_{ie}	Small signal input impedance, output short circuit
h_{oe}	Small signal output conductance, input open circuit
h_{fe}	Small signal current gain
Tol	Percentage tolerance of h_{fe} value
f_T	Turnover frequency in MHz

All of these parameters are for common-emitter connection – the necessary adjustments will be made by Aciran if you use a circuit in which a transistor is connected in common-base or common-collector mode.

The real problem is to find these quantities for the transistors that you are likely to use. Few catalogues print all of the h-parameters, and though h_{fe} and f_T are usually easy to find, it can be quite a struggle to obtain the h_{ie} and h_{oe} figures. The units are normally k for h_{ie} and µs (micro Siemens) for h_{oe}.

If you want to make use of an unlisted transistor you must know values for all of these parameters, and fill them in when the form appears. This form will appear automatically if you answer N to the question about using Models – note that the answer will be assumed to remain as N until you alter it to Y.

It is much more useful to add all of the transistors you are likely to use into the Models file, and to do so you need only a text editor or a word processor. If you use a word processor it must be set so as to produce what is called an ASCII file, meaning a file of characters with no invisible codes. All word processors will produce such files, but you need to consult the manuals to find out how to make them do so. WordStar, for example, requires you to select a Non-document file at the time when you open a file. WordPerfect uses the opposite system of allowing you to type the file, and then opt for saving it as a DOS file. If you are in any doubt about using a word processor, there are several simple low-cost text editors available from the Public Domain Software Library.

The rules for creating transistor model files are:

1 Each file consists of one quantity on each line and requires six lines of typing.

2 No quantities must be omitted, and the quantities must be typed in the correct order.

3 The ENTER key must be pressed to end the entry of each quantity.

4 The order of entry, which must be adhered to, is:

>Name
>h_{ie} value
>h_{oe} value
>h_{fe} value
>Tolerance of h_{fe}
>f_T value

Following entry of the f_T value, the file can be saved, using the transistor name for the name of the file and the extension letters of TRN.

For example, the BC107 model is filed as BC107.TRN and consists of the lines:

```
BC107
2.7k
18u
192
35
300M
```

describing the parameters in order.

FETs

Aciran deals with FETs as it deals with bipolar transistors by creating files for the various types or by filling in details on a form. The use of an FET in a circuit does not require you to specify the DC current, so that the analysis of FET circuits is in this respect easier than that for bipolar transistors. The set that is provided is:

- 2N3819 N-channel general-purpose amplifier
- 2N3821 General-purpose amplifier
- 2N4338
- 2N4393 N-channel switching
- 2N4416 N-channel VHF/UHF amplifier
- 2N4417A
- 3N163
- BF256LA
- J300
- MPF102 N channel VHF
- U320

Finding characteristics for FETs that are not in the set provided along with Aciran is not easy however. FET characteristics are entered into a file or into a form as required, and consist of:

- The FET designation
- The g_m (or Y_{fs}) value
- Tolerance of g_m value
- Capacitance, drain to source
- Capacitance, gate to drain

Most tables will provide g_m or Y_{fs} values, but very few print the inter-electrode capacitance values and you may have to guess these for some FET types unless you can find design information from the manufacturers. The capacitances affect mainly the higher-frequency uses, but their values must be present in order for Aciran to use the FET model. Note that these capacitances are the only frequency-limiting factors in the FET model; there is nothing corresponding to f_T or to gain-bandwidth product in the Aciran model of an FET.

A tone control example

The Baxandall type of tone control has always been a popular circuit both with professional audio designers and with amateur enthusiasts, but it is more often copied than designed from scratch, and it is by no means simple to analyse in detail. A modern version using an FET and a bipolar transistor combination is a good way to illustrate the analysis of FET circuits by Aciran ; the methods used are so similar to analysis of bipolar transistor circuits that further examples are not really needed. The problem of modelling this circuit with Aciran is to account for the use of potentiometers, and these have to be modelled by using two-resistor chains, with one set of readings taken in each of three possible settings. In other words, the connection will be made to the centre point of the resistors, to one end, and then to the other end in three separate runs.

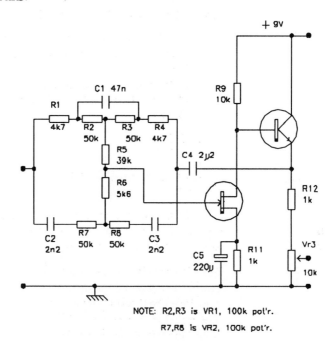

Figure 4.32 *The classic Baxandall tone control circuit, here using one FET and one bipolar transistor. The bass and treble potentiometers are represented by resistor pairs.*

The circuit is shown in Figure 4.32, with the tone potentiometers shown as separate resistors. The resistors R_2 and R_3 represent a 100k potentiometer (bass control), and resistors R_7 and R_8 similarly represent the 100k treble control potentiometer. In the Aciran representation, the circuit can be analysed as shown (with the source of the FET connected to Node 0), and then with R_2 set at 100k and R_3 at 1 ohm to represent one extreme setting of the bass potentiometer. Another run with R_2 at 1 ohm and R_3 at 100k can be made to represent the other extreme setting. The volume control Vr_3 is taken as a 10k fixed resistor for the purpose of analysis.

The analysis with controls in mid-position as shown in the diagram provides the amplitude graph of Figure 4.33. At first sight, this does not look like a flat-frequency response, but this is because of the scale which Aciran is using – there is only 2.06 dB of difference between the top of the graph and the bottom, so that the maximum variation in signal amplitude over the range of 10 Hz to 20 kHz is less than 0.4 dB. To all intents and purposes, then, this is a graph of a flat-frequency response.

The next step is to set resistors R_2 and R_3 so as to simulate the maximum bass boost by making the value of R_2 1 ohm and that of R_3

124 USING ACTIVE CIRCUITS

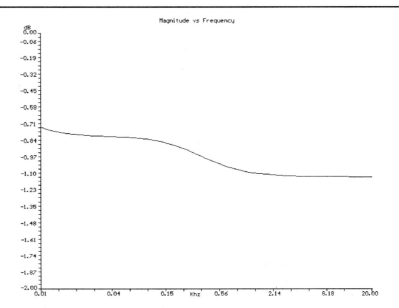

Figure 4.33 *The flat response of the Baxandall circuit when all controls are set level.*

100k. Note that Aciran will change an entry of 1 into 1000m so that checking the value of R_2 can sometimes make you wonder if you carried out the alteration correctly unless you read the whole of the entry. The result of using this position is shown in Figure 4.34, showing a 25 dB boost for the lowest frequencies as compared to the higher frequencies. The boost is concentrated below 150 Hz, in the true bass region, rather than extending from mid-frequencies as occurs in many tone control circuits.

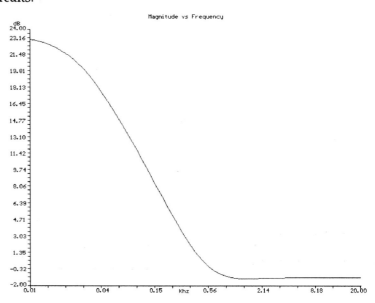

Figure 4.34 *Maximum bass boost settings show that the boost extends to very low frequencies without affecting mid-frequencies.*

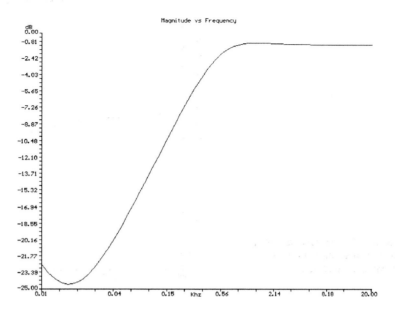

Figure 4.35 *Maximum bass cut has a similar effect, but with an odd turnover at a very low frequency.*

If the values of R_2 and R_3 are now reversed to produce the effect of maximum bass cut, the result is as shown in Figure 4.35. There is a fall of about 25 dB at around 25 Hz, though there is an odd bending of the graph at the lowest frequencies. As before, the roll-off frequency is around 150 Hz, and the rate of decrease is very satisfactory.

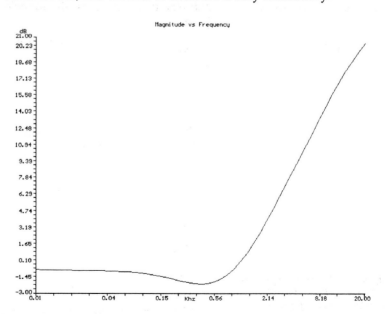

Figure 4.36 *Maximum treble boost extends to the limits of this analysis (20 kHz), with a slight trough at 500 Hz.*

126 USING ACTIVE CIRCUITS

Restoring the value of R$_2$ and R$_3$ to 50k each allows the effect of the treble controls to be checked by altering the values of R$_7$ and R$_8$. Starting with values of R$_7$=1Ω and R$_8$=100k provides the maximum treble boost position as shown in Figure 4.36. The turnover frequency is around 2 kHz, but note the slight dip in response around 500 Hz. With the values of R$_7$ and R$_8$ reversed, the response is as shown in Figure 4.37, almost a mirror image of the previous graph but with a cut of more than 25 dB. In this graph again, there is a slight hump at around 500 Hz.

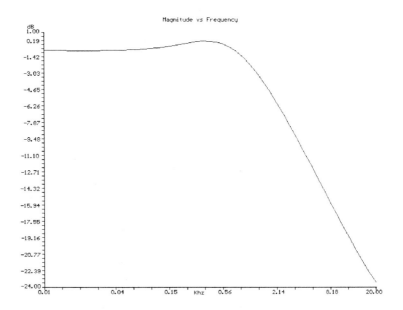

Figure 4.37 *Maximum treble cut follows a mirror image pattern, even to the extent of the bump at 500 Hz.*

Response graphs for tone control circuits usually show composite pictures with these views superimposed. This cannot be done using Aciran because even if the graphs are gathered as graphics files their dB scales are all different and the graphs are not plotted with the mid-frequency region placed on a 0 dB line. If you want to prepare composites from Aciran graphs, then, you have to start with the shape of the 'flat' response drawn along a 0 dB line, and then plot the other responses relative to this line. There is no easy automatic way of carrying out such work.

5 Operational amplifiers

Of all the active devices that Aciran can deal with, the OPamp set is in many ways the easiest to use because there are few parameters and if new devices need to be added to the list in the MODELS set the characteristics are easier to find than those of bipolar transistors or FETs. The existing set consists of the following types:

    LM124 1 MHz GB product
    MC1558 1 MHz GB product
    NE530 3 MHz GB product
    NE538 6 MHz GB product
    NE5512 3 MHz GB product
    NE5532 1.2W max., 9 V/µs slew rate, 10 MHz GB product
    NE5534 800 max., 13 V/µs slew rate, 10 MHz GB product
    TL084 FET 680 mW max., 13 V/µs slew rate, 10 MHz GB product
    UA741 500 mW max., 0.5 V/µs slew rate, 1 MHz GB product
    UA747 670 mW max., 0.5 V/µs slew rate, 1 MHz GB product
    Standard (a cover-all set of parameters, 1 MHz GB product)

To add to the set, the following must be specified:

Name
Input impedance
Output impedance
Open loop gain
Gain bandwidth product (GB)
Open loop gain tolerance %

Of these, the most difficult to find are usually the input impedance and the output impedance. Note that the model used in Aciran does *not* require slew-rate information, because slew rate is a limitation that applies more to non-linear signals and to large amplitude inputs, but the slew-rate limitations of operational amplifiers should be remembered if you are concerned with high-frequency waveforms at the higher amplitude levels.

The Standard model is:

Input impedance	100M
Output impedance	600R
Open loop gain	100 dB
Gain bandwidth	1 MHz
Gain tolerance	50%

This can be used as a general-purpose OPamp in circuits where a device is not specified. These parameters can also be used as a guide for a device you want to model but for which insufficient data is available. The output impedance for OPamps is generally in the range 15R to 600R, and the input impedance is seldom lower than 100M. The open loop gain is also seldom less than 100 dB, so that for most OPamps the only factor that needs to be specified closely is the gain bandwidth, also known as the unity gain frequency. If you know the gain bandwidth for an OPamp, you can make intelligent guesses on the other parameters, and the quantities such as 100M for output impedance and 600R for input impedance are usually suitable.

Where a circuit to be analysed makes use of an OPamp, only three node connections to the OPamp need be specified. These are the nodes for the *+input*, the *–input* and the *output* respectively. Any connections used for power supply or for offset adjustment are ignored because these are not signal connections, and Aciran, as always, deals only with signals. Specialised audio amplifier ICs with other signal connections cannot be analysed by Aciran unless you can obtain an equivalent circuit (using OPamps or other standard components).

The range of circuits you are likely to encounter for analysis becomes considerably greater when these operational amplifiers are added to the list of active devices, since more circuits nowadays use operational (or similar IC) amplifiers than use discrete bipolar or FET devices. In addition, circuits that use OPamps do not require you to specify precise DC bias conditions, making it considerably easier to analyse such circuits. Since OPamps (or a few specialised audio amplifiers which can be added to the OPamp models) are used in many circuits for which passive components determine the characteristics almost entirely, this chapter will show several examples of circuits whose analysis is time-consuming by conventional methods, and which can be used in your own circuit designs. We shall start, though, with the conventional circuits that appear in all texts dealing with OPamps, so that these well-known examples can be checked against the Aciran analysis.

☐ The more specialised audio (and other linear) ICs can be modelled using the constant current and constant voltage generators, provided that you can obtain an equivalent circuit for the IC.

Simple inverting amplifier

The simple inverting amplifier circuit of Figure 5.1 is the most common and most fundamental OPamp circuit. The input is taken by way of a resistor which determines input impedance to the inverting input, and feedback is also applied to the same point from the output through another resistor. For frequencies well within the GB limit, the gain is determined by the ratio of resistor values R_2/R_1. Stray capacitances, marked as C_{s1} and C_{s2}, have also been shown in this drawing. They will not be shown in other drawings in this chapter, because if we believe the simple feedback theory they have little effect on most OPamp circuits since both points are at a low impedance (the input because of the feedback).

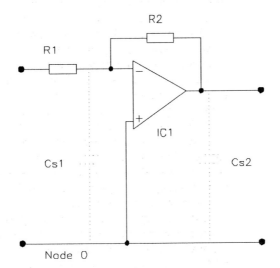

Figure 5.1 *The simple OPamp inverting amplifier, in which the gain is determined by the R2/R1 ratio. Stray capacitances have been dotted in.*

The circuit can be tried in Aciran to see how well analysis accords with the simple theory, and in particular to check how the GB product affects the gain. Using values of $R_1 = 10k$, $R_2 = 470k$ with the STANDARD model of OPamp gives a gain of 47 times, 33.44 dB, and the bandwidth should be 1M/47, around 22 kHz. The analysis is illustrated in Figure 5.2. This has been swept between 10 Hz and 1 MHz, and it shows that the gain is exactly as calculated, but that the frequency response curve is not like anything mentioned in the text books, with a very large peak at the turnover frequency. This is accompanied by an equally large phase change at the same frequency, Figure 5.3.

Looking at the input and output impedances (not illustrated here) reveals similar effects. The input impedance remains around the 10k mark for the lower frequencies, but dips down to 414 ohms at around 20 kHz, then rises to over 40k and dips again to 414 ohms. The output impedance maintains a value of a fraction of an ohm for a larger range

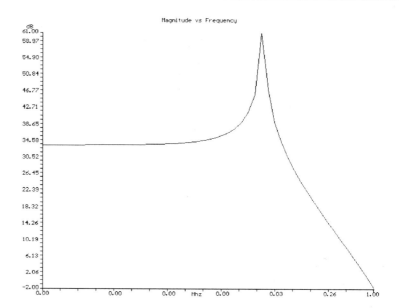

Figure 5.2 *The frequency response plot for the OPamp inverter. The presence of stray capacitances has a profound effect on the shape of this graph.*

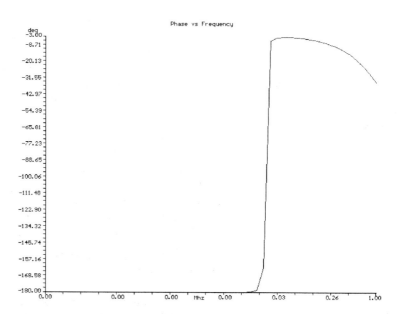

Figure 5.3 *The phase diagram for the OPamp inverter, showing the violent changes at the turnover frequency.*

to around 80 kHz before dipping to an indicated negative value and then rising to a few hundred ohms.

When the stray capacitance components are deleted, the picture changes. The graph of gain plotted against frequency is much the same, but the phase graph is considerably altered (Figure 5.4), with a graph

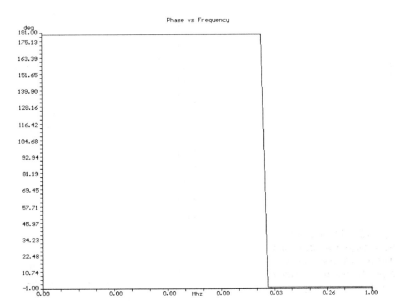

Figure 5.4 *The text-book response that is obtained when stray capacitances are ignored.*

that looks as if it had been made using a ruler. The phase is almost constant at 180° up to the turnover frequency, when it abruptly changes to −1° and remains constant at this value (so that the inverting input is no longer inverting but in phase). The input impedance remains at around 10k to the turnover frequency and then rises sharply. The output impedance is negligibly low up to the turnover frequency and then rises violently.

☐ These values of input and output impedance should be read from the tables rather than from the graph, because the changes are so large that the graph is misleading because of the large scale range. In particular the input impedance graph will show a value of around 400 ohms for the lower-frequency ranges, because of the very cramped scale that is used with the next value shown as 430783.97 ohms. If in doubt, always take readings from the table.

The moral here is that stray capacitances have a considerable effect on any OPamp circuit, particularly near the point where the GB frequency causes its violent phase shifts. Internally, any OPamp is a very complicated device, and the GB turnover is not simply a matter of a gentle slope such as would be caused by an RC time constant, it is the cumulative effect of the turnover frequencies of a large number of identical transistors in IC form. OPamp circuits should be used well within their frequency limits, and filtering may have to be used to ensure that frequencies near the GB limit do not appear at the input. The output impedance, incidentally, appears as 624 ohms for most of

132 OPERATIONAL AMPLIFIERS

the range when realistic drive and load impedances such as 10k are specified.

☐ Always check that your OPamp circuits are being operated between sensible values of drive and load impedances, as these can make a considerable difference to the way that feedback loops operate. In particular, do not leave the load impedance at 100M if you want to see sensible values of output impedances read from the tables. The other standard circuit (ignoring differential amplifier circuits) is the follower type in which the +*input* is used for the signal and the −*input* for feedback (Figure 5.5). The low-frequency gain for this configuration is given by:

$$\frac{R_2 + R_4}{R_2}$$

and the resistor R_3 is used only to ensure an earth return for the input. We can look at an example in which $R_1 = 10k$, $R_2 = 47k$, $R_3 = 1M$ and $R_4 = 10k$.

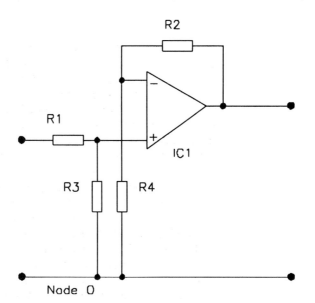

Figure 5.5 *The non-inverting or follower-with-gain type of OPamp circuit.*

Using Aciran, the low-frequency gain is shown as 15 dB (15.032 to be precise) which agrees perfectly with the simple theory, and the phase reversal at the output is at a frequency just below 200 kHz, which seems reasonable for a gain bandwidth factor of 1 MHz with a voltage gain of around 5.7 times. The graph, Figure 5.6, shows the same peak as was observed for the previous example. This is to be expected, as it is a consequence of the internal construction of the OPamp rather than of

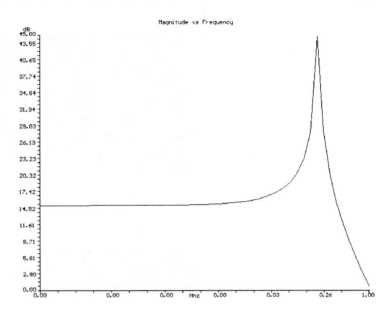

Figure 5.6 *The frequency response graph for the non-inverting circuit.*

the circuit itself. The peak is at a considerably higher frequency this time because the voltage gain of the circuit is lower. The input impedance of the circuit is high, just over 1M, for the useful frequency range, and the fluctuations around the turnover frequency are not so violent as in the previous circuit. The output impedance remains very low for the useful range and rises to a peak of just over 6k around the turnover frequency, falling again at higher frequencies.

The follower circuit is useful as an impedance converter, but there is a limitation to its use that Aciran cannot point out. Using an OPamp in this mode is valid for small-amplitude signals only because there is a limit to the amplitude that can be applied at the +input terminal, set by the value of maximum common mode input voltage (voltage applied to both inputs, as happens when the *+input* is used along with feedback to the *–input*). This is not a limitation that will affect many applications, because the maximum common-mode input voltage is usually quite large (of the order of 15 V for a 741), but it must not be forgotten.

When the standard 10 pF values of strays are added to this circuit, using three capacitors (one at each input and one at the output node), the effects on the circuit can be analysed. As before, the gain graph is virtually unaltered, but the phase graph is smoothed out, with a less-violent change and a turnover at several hundred kHz. The input impedance is considerably affected, starting to dip from its 1M value at just over 10 kHz and reducing to almost zero at just over 30 kHz. The output impedance is better behaved, remaining low for frequencies below 30 kHz and reaching a peak of around 3k for a frequency around 250 kHz.

☐ Once again, this shows how the simple equations for using an OPamp can be misleading in the presence of quite reasonable amounts of stray capacitance. It is normal, for example, to assume that the input impedance of a follower circuit will remain high over the whole usable frequency range.

The circuit without stray capacitances will behave rather differently when a different OPamp is substituted. Use the *Modify* menu to select the *OPamp*, and substitute the NE5534 for the STANDARD model. The NE5534 has a GB factor of 10 MHz, and the effect of the substitution will be to raise the bandwidth of the gain and phase graphs. The input and output impedances, in the absence of stray capacitances, will also be maintained to higher frequencies. In the presence of strays, however, the input impedance in particular will still change significantly at quite low frequencies, and this should be borne in mind when the follower type of circuit is used.

Simple filter networks

The fact that an OPamp circuit can be used as a theoretically perfect feedback circuit over a limited frequency range makes such circuit useful for wave-shaping applications. In particular, the type of filters known as *active filters* (see later) can be used to replace older designs for applications in which the limited bandwidth of the OPamp circuit is not a disadvantage. Before we look formally at active filter circuits, though, it is useful to try out a pair of simpler circuits, the basic integrator and differentiator circuits.

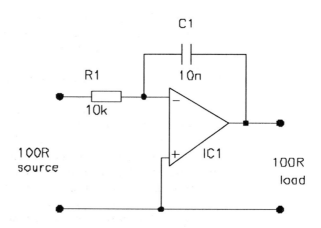

Figure 5.7 *An OPamp integrating circuit of the conventional type.*

Figure 5.7 shows an integrating circuit based on the STANDARD OPamp model, and using the combination of 10k and 10 nF as a time constant, with drive and load impedances of 100 ohms each. The phrase 'integrating circuit' implies a square wave input, but as a linear circuit we can analyse only the response to a range of sine waves (though some analytical circuits can analyse the response to a square input). The analysis will be done over the range 10 Hz to 100 kHz, though for the STANDARD model of OPamp this is slightly on the generous side.

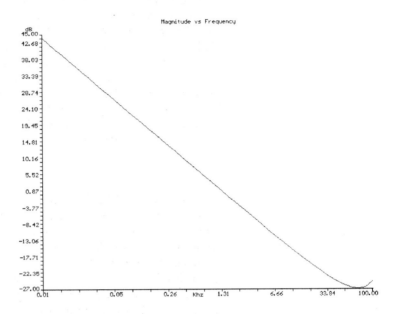

Figure 5.8 *The amplitude frequency plot for the integrator, showing the expected behaviour.*

The plotted response is shown in Figure 5.8. The graph of amplitude plotted against (log) frequency is almost a straight-line graph, as would be expected, but its linear region does not extend to 100 kHz, so that some care would be needed when using this circuit to ensure that its frequency limits were not approached. The total gain range is more than 70 dB, from around +44 dB at low frequencies to –27 dB at the highest useful frequency of about 69 kHz. The input impedance is 10 kHz for low frequencies, but the phase of the input impedance jumps abruptly from –179° to –1° at a frequency of 25 Hz and remains at this level for the remainder of the frequency range. The output impedance is low, of the order of 1 ohm, until about 6 kHz, after which it climbs steadily to a value of around 70 ohms at 100 kHz.

The opposite form is the differentiator, whose basic circuit is shown in Figure 5.9. The same drive and load impedances have been used, and the time constant is also the same as that of the previous example. Remember that a differentiating circuit in a feedback loop performs the

136 OPERATIONAL AMPLIFIERS

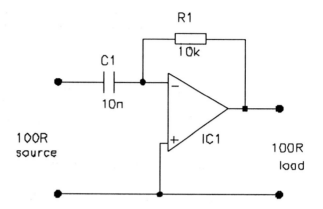

Figure 5.9 *The basic differentiator circuit using an OPamp.*

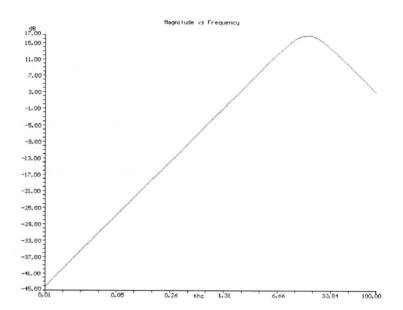

Figure 5.10 *The amplitude frequency graph for the differentiator circuit.*

action of an integrator in a straightforward circuit, just in case you have doubts about this and the preceding circuit.

The amplitude response is illustrated in Figure 5.10, with its gain rising from a value of almost –45 dB at 10 Hz to about +17 dB at 16 kHz. Note that this is not simply a mirror image of the previous example as you might have thought. Both the decibel ranges and the frequency turnover are quite different for this circuit despite the use of identical time constants, showing that nothing should be taken from granted when active circuits with feedback are being considered. The input impedance magnitude starts very high, well over 1.5M, but reduces to around 1 ohm at 1 kHz or so; its phase changes violently at 6.66 kHz.

The output impedance is very low up to a frequency of about 6.66 kHz and then rises to a few hundred ohms.

Other circuits

The application of OPamp feedback circuits covers virtually all types of circuitry which formerly was used in passive form only, and of the many circuits that can be used, the Twin-T is one whose OPamp form is particularly interesting because of the interaction between the passive circuit and the OPamp. A circuit is shown in Figure 5.11, using 22k and 10n as the main time constants, giving a predicted notch frequency of 723 Hz. Since the Twin-T is being used in a feedback loop, it will cause the overall response to peak at this frequency.

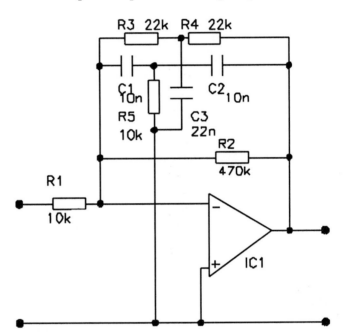

Figure 5.11 *A Twin-T circuit constructed using an OPamp.*

The Aciran output graph, Figure 5.12, shows the expected peak, and the table shows that the frequency is around 725 Hz. The value of peak gain is 38 dB, slightly more than would be expected from the use of 470k as R_2, but limited by this value. If the value of R_2 is reduced, the size of the peak relative to the gain at other frequencies will be reduced. The phase graph, Figure 5.13, shows the expected sharp change at the critical frequency and there are also sharp changes in input impedance and output impedance around the same frequency. A larger peak can be obtained by using smaller values of R_3, R_4 and R_5 and correspondingly larger values of C_1, C_2 and C_3 (to maintain the same time con-

138 OPERATIONAL AMPLIFIERS

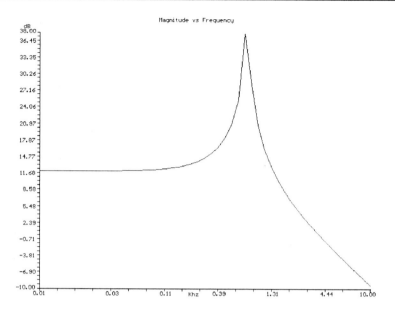

Figure 5.12 *The output graph for the Twin-T circuit.*

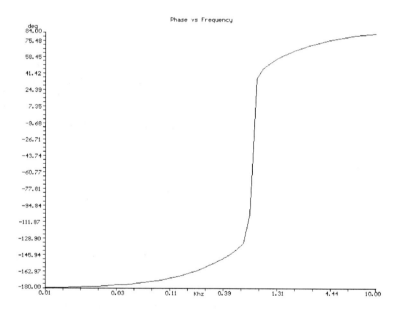

Figure 5.13 *The phase frequency plot for the OPamp Twin-T circuit.*

stants). Note that the shape of the graph is not symmetrical – the response does not return to the value of 6 dB following the peak.

A much more symmetrical response is obtained when a Wein Bridge is used as the feedback circuit (Figure 5.14). This circuit shows the Wein network in the input and feedback portions of the circuit, and also

OTHER CIRCUITS 139

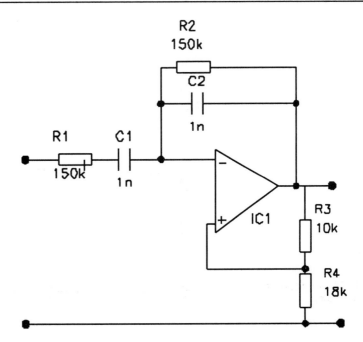

Figure 5.14 *A Wein Bridge circuit constructed using an OPamp.*

shows a less common feature, the use of some positive feedback to increase the gain of the circuit. The positive feedback is applied by a potential divider R3, R4 to the +input, and it is important in this circuit to ensure that the amount of positive feedback is limited to less than the amount which would cause instability. In this case stability is assured if the value of R4/R3 is less than 2; the value used is 1.8. The addition of positive feedback has a considerable effect on the equivalent Q-factor for the network, which is given by:

$$\frac{1}{2 - \frac{R_4}{R_3}}$$

The closer the ratio of R4/R3 can approach the value of 2, the higher the equivalent Q of the network.

The Aciran graph for amplitude plotted against frequency, using the values shown in Figure 5.14, is illustrated in Figure 5.15. The peak is at 1 kHz and the attenuation on each side of the peak is about 17 dB per octave for the values illustrated. This value will depend heavily on the R4/R3 ratio, but since the positive feedback reduces stability it is undesirable to depend too much on working with a ratio that approached the critical value very closely.

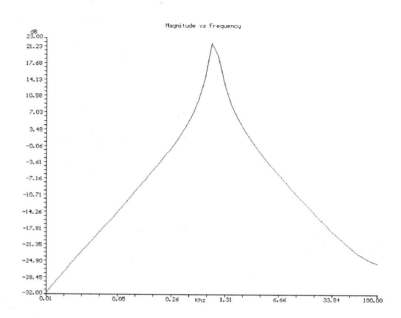

Figure 5.15 *The plotted response for the Wein Bridge circuit.*

Sallen & Key active filter circuits

Active filters using capacitors and resistors have to a considerable extent replaced the older type of LC filters for most circuit applications in the lower frequency ranges. The main disadvantage of simple RC filter circuits is that they have low Q-factors, but this can be overcome either by using a bridge circuit which is close to its balance condition, or by using the filter components in a feedback circuit. In either case, active devices, which in practice are almost always operational amplifiers, are an essential part of the complete circuit. Aciran can analyse either type of active filter, but cannot be used for another type, the switch-mode type, in which several different filters are consecutively switched into a signal path at a rate much higher than the fundamental frequency of the signal.

The bridge type of filter has already been covered by earlier comments on the Wein Bridge and Twin-T circuits, so that this section deals with the most common of the other type of active filter, the Sallen & Key type, named after R.P. Sallen and E.L. Key whose article in *I.R.E. Trans.* in 1955 launched this type of filter circuit. These active filters can be obtained in low-pass, high-pass or bandpass form.

Figure 5.16 shows a simple low-pass circuit using an integrator in the input and a differentiator in the feedback path. This is the normal arrangement of a S&K filter, using one form of time constant in the direct path and the other form in the feedback path. In this type of filter,

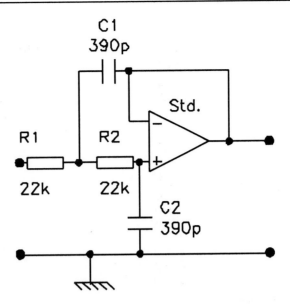

Figure 5.16 *A low-pass Sallen & Key type of active filter circuit.*

the resistor values are usually equal, and the capacitor values are $C_1 = 2C_2$, so that the theoretical response can be found from the equation:

$$f = \frac{1}{2\sqrt{2}\,\pi R C_2}$$

assuming that $R = R_1 = R_2$ and $C_1 = 2C_2$.

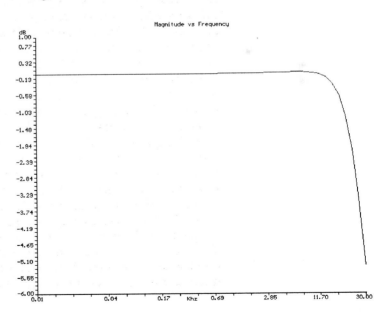

Figure 5.17 *The analysed response for the low-pass circuit.*

The graph of Figure 5.17 shows the very level response over the passband and also the steep slope of the cut-off. The 3 dB point read from the analysis is around 22 kHz. Changing the ratio of R_1 to R_2 and C_1 to C_2 will alter both the 3 dB frequency and the rate of cut, and Aciran can be used to investigate these variations. The phase-shift graph, Figure 5.18, is smoothly changing, with no abrupt steps of value.

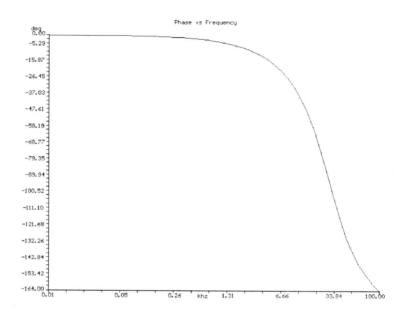

Figure 5.18 *The corresponding phase-shift graph, with no sudden changes evident.*

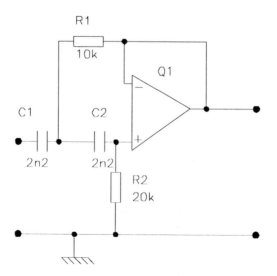

Figure 5.19 *A high-pass Sallen & Key filter.*

A high-pass filter takes the expected form of Figure 5.19, but this time the capacitors are equal and the resistors are in the ratio $R_2 = 2R_1$. This makes the formula take the same form as before, and we can test the performance with the quantities shown in the circuit diagram. The graph appears in Figure 5.20 showing the high-pass characteristic. A listing of the amplitudes shows that the 3 dB point is the expected 5 kHz in this example. Once again, the phase graph, Figure 5.21, is smoothly changing with no abrupt steps.

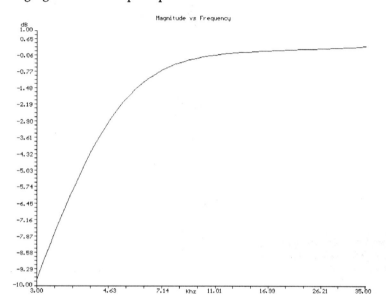

Figure 5.20 *The amplitude response for the high-pass circuit.*

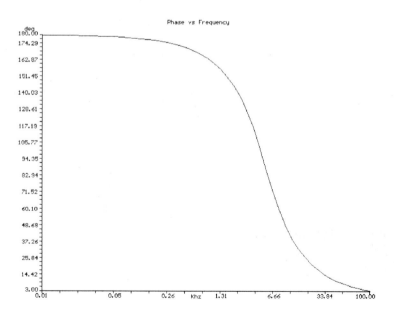

Figure 5.21 *The phase response for the high-pass active filter.*

A further stage of filtering can be added to the circuit, using a series capacitor of 2n2 at the output, with a 20k resistor to earth. This alters the turnover frequency to 6 kHz and makes the 1 kHz response some –40 db rather than the –29 dB of the original circuit (Figure 5.22). This amount of filtering can be very effective, and the sharp changes of phase (Figure 5.23) compare with those found using LC circuits.

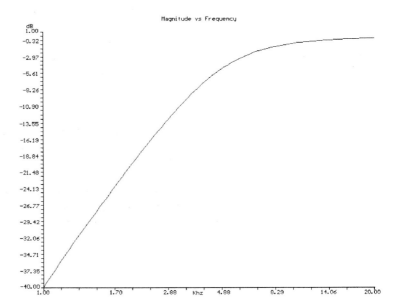

Figure 5.22 *The effect on the amplitude graph of adding a further simple stage of filtering.*

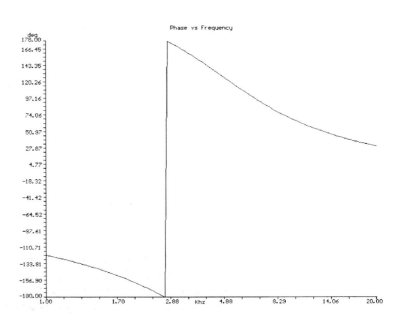

Figure 5.23 *The corresponding phase graph for the enhanced filter circuit.*

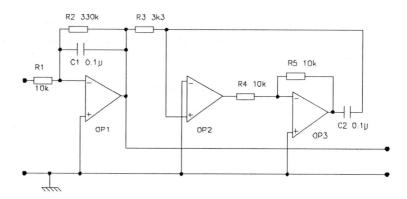

Figure 5.24 *An OPamp triplet circuit with remarkable characteristics.*

Sets of three OPamps, with both differentiating and integrating actions, can form the basis of circuits which have quite remarkable characteristics. Figure 5.24 shows one such circuit in which the values indicated produce an amplitude phase graph which, on the decibel and log frequency scales (10 Hz to 2 kHz) is almost a straight line, Figure 5.25. The same characteristic shape is maintained if the frequency range is extended to 20 kHz, with an attenuation of −42 dB at 20 kHz.

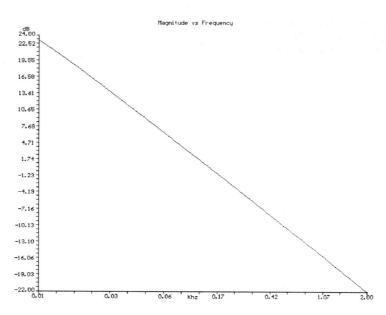

Figure 5.25 *The virtually straight slope of the low-pass characteristic of the OPamp triplet.*

Using the same circuit, altering the capacitor values to 100 pF makes the characteristic become a steep-cut low-pass one, Figure 5.26, and reducing the capacitance still further creates a distinct hump at around 8 kHz with a sharp fall at higher frequencies.

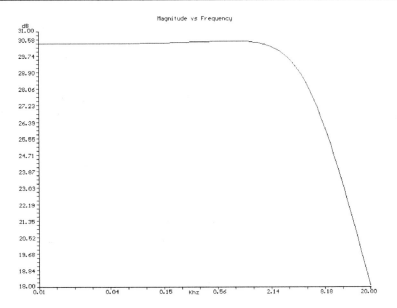

Figure 5.26 *The effect of altering the capacitor values to make the low-pass action extend over a band of frequencies.*

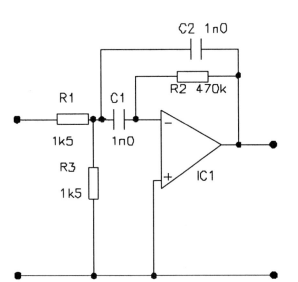

Figure 5.27 *Another form of bridge circuit used with an OPamp.*

Active circuits can also be used to create bandpass and notch filters. We have already seen the Twin-T and Wein Bridge circuits in use, but the circuit in Figure 5.27 is one that also uses a bridge form of structure with two feedback loops. The effect is that of a bandpass filter with a fairly narrow bandwidth for audio frequencies. The central frequency can be determined by altering the value of R_3 without altering the other values. The design equation is:

$$f_0 = \frac{1}{2\pi C} \sqrt{\frac{(R_1 + R_3)}{R_1 R_2 R_3}}$$

with $C = C_1 = C_2$. With the values shown, this predicts maximum gain at 8.486 kHz. The equation for equivalent Q-factor is:

$$Q = \frac{1}{2} \sqrt{\frac{R_2 (R_1 + R_3)}{R_1 R_3}}$$

and for this circuit ought to be of the order of 12.

The Aciran analysis provides a gain frequency graph as shown in Figure 5.28. The graph is fairly symmetrical and is sharply-peaked, providing good selectivity at this comparatively low frequency. The phase-shift graph of Figure 5.29 shows the sharp change that is typical of a circuit with a reasonably high Q-value. The circuit of Figure 5.30 uses OP2 to simulate an inductor which will resonate with C_1 to produce a notch response whose notch frequency and equivalent Q depend on the circuit values.

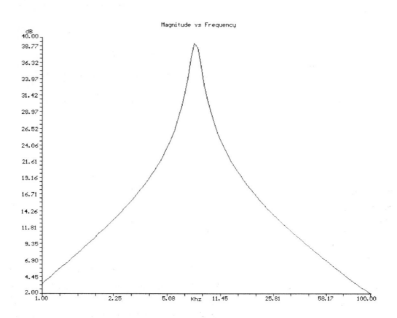

Figure 5.28 *The bandpass response for the OPamp bridge circuit.*

148 OPERATIONAL AMPLIFIERS

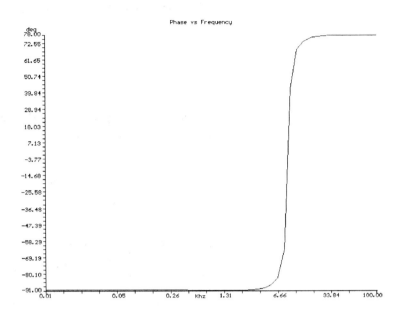

Figure 5.29 *The phase response for the OPamp bridge circuit.*

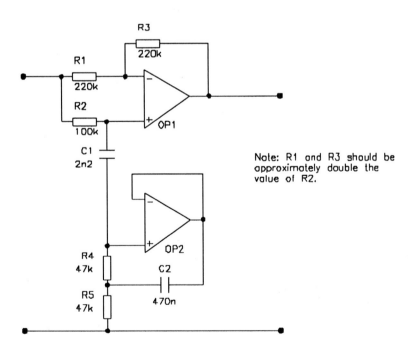

Note: R1 and R3 should be approximately double the value of R2.

Figure 5.30 *A circuit in which one OPamp simulates an inductor.*

The frequency is determined by:

$$\frac{1}{2\pi\sqrt{C_1 C_2 R_4 R_5}}$$

which for the component sizes illustrated should give a notch frequency of around 105 Hz. The equivalent Q is found from:

$$Q = \frac{1}{R_4 R_5}\sqrt{\frac{C_2 R_4 R_5}{C_1}}$$

giving for this example a value of around 7.

The values illustrated produce the amplitude/frequency graph of Figure 5.31 and the table shows that the notch frequency is 105 Hz as theory predicts. The phase-shift graph shows the expected sharp reversal at the notch frequency, Figure 5.32. Looking at the formula, the simplest way of increasing the depth of the notch would appear to be by increasing the ratio of C_2 to C_1, but trying this on the Aciran model shows little useful increase.

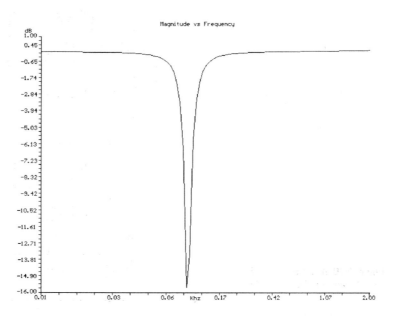

Figure 5.31 *The frequency response of the circuit, showing the notch response.*

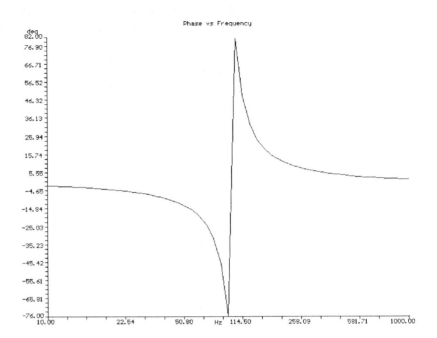

Figure 5.32 *The corresponding phase response for the notch filter circuit.*

Audio circuits

The active filter circuits that have already been considered are used to a considerable extent in audio circuitry, where such items as scratch and rumble filters are required for pre-CD audio. The circuit analyses that follow are for more specialised audio circuits of the type that have been used in the LP days. Now that the supplies of LPs are drying up, these circuits are of less pressing interest, but the methods that they use are applicable to other systems, and in any case, there are still millions of LPs which will have to be preserved by transcribing them to tape (if such a step ever becomes totally legal).

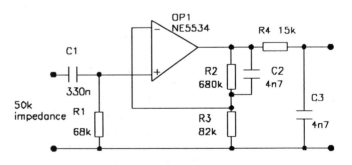

Figure 5.33 *A well-tried OPamp magnetic cartridge input stage.*

Figure 5.33 is another magnetic cartridge input stage, using the high-performance NE5534 OPamp in this case. The time constants are calculated to provide the RIAA playback characteristic to fairly close limits (the value of R3 can be trimmed to a slightly lower value with a 1M resistor to make the response closer to RIAA). The use of the NE5534 avoids any problems of GB factor of the type which have been noted earlier.

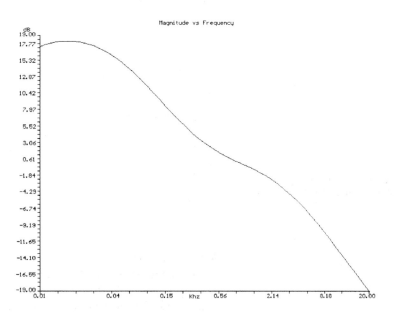

Figure 5.34 *The amplitude response for the magnetic cartridge input stage.*

The analysis of amplitude/frequency is illustrated in Figure 5.34. The shape of the graph is correct for the RIAA characteristic, but the table of quantities should be used to check that the response is within reasonable limits (1 dB is reasonable). This checking is not easy, because tables of correct RIAA responses are not easy to come by, and never were, and the graphs which are printed in books are on too small a scale to be useful for checking.

Preamplifier stages for ceramic pickups were used to a considerable extent at one time, and the circuits are useful for any application in which the driving impedance is very high. This was a difficult task to achieve in the early days of transistor circuitry, but it became much simpler when FETs and OPamps became available with their naturally high-input impedance levels.

Figure 5.35 illustrates a typical circuit for this purpose, carrying out the impedance matching (at least 1M input impedance was called for) and the required equalisation. The high-input impedance is achieved by using positive feedback in a bootstrap circuit, and the capacitor C_2 should be a low-leakage tantalum type. The capacitor C_1 is not part of the circuit, but the equivalent capacitance of the ceramic cartridge.

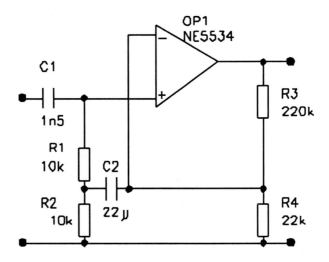

Figure 5.35 *An input stage with high impedance, originally used for ceramic pickup cartridges.*

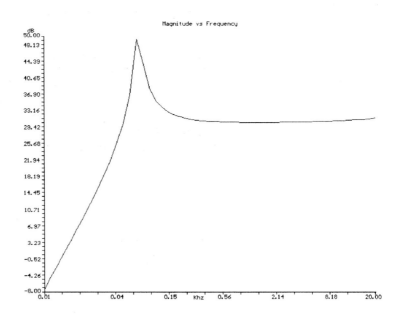

Figure 5.36 *The frequency response for the ceramic cartridge input circuit.*

The amplitude graph, Figure 5.36, is a rather curious one. The response below 40 Hz is poor because of the cartridge capacitance C_1, but above 150 Hz the graph is fairly flat. The peak at around 85 Hz constitutes a large bass boost of some 20 dB, caused by the cartridge capacitance resonating with the inductive input impedance of the circuit at this frequency. The amplitude of input impedance is high over the audio range, but the phase of input impedance changes violently (Figure 5.37) due to the simulated inductive impedance at low frequencies and its resonance with the cartridge capacitance. To see the perfor-

AUDIO CIRCUITS   153

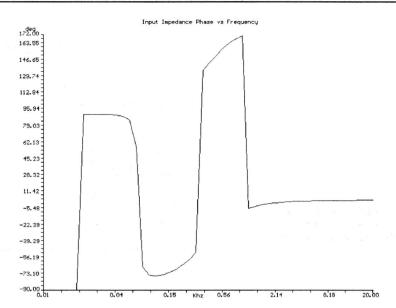

Figure 5.37 *The unpleasant shape of the phase characteristic for the circuit.*

mance of the circuit in the absence of the cartridge capacitance, delete C_1 from the circuit and analyse again.

The third classical audio preamplifier circuit is the tape preamplifier. Old designs were intended for use with open-reel tape, so that their time constants are generally unsuited to cassette uses. The circuit in Figure 5.38 provides the required equalisation for normal ferric tape (120 µs), with the bass time constant of 3180 µs.

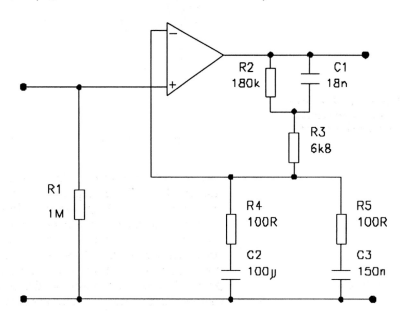

Figure 5.38 *A tape input circuit with equalisation components.*

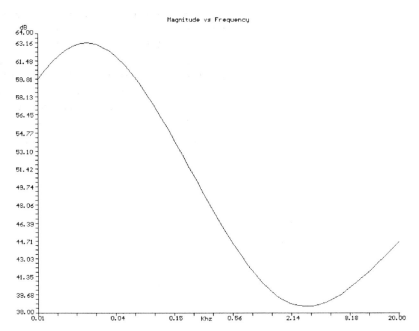

Figure 5.39 *The amplitude response of the tape input circuit.*

The response is illustrated in Figure 5.39, with the bass boost a maximum at around 29 Hz and the minimum gain at around 2.77 kHz. The roll-off from the slope is determined by the 120 µs time constant and is not so easy to estimate.

Finally, look at two high-slope filter circuits due to John Linsley Hood and using bootstrap methods with OPamps. The high-pass filter of Figure 5.40 is designed to have a turnover of 30 Hz and a slope of 20 dB per octave below this so as to act as an audio rumble filter. The circuit

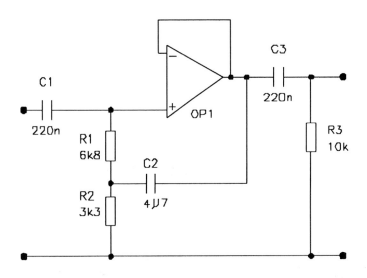

Figure 5.40 *A high-pass filter circuit designed by John Linsley Hood.*

is operating with unity gain so that any OPamp is suitable from the bandwidth point of view, but for audio applications it is always better to choose an OPamp with a good frequency range, low noise, and high slew rate. The NE5534 has been used in the examples once again. The Aciran analysis of Figure 5.41 shows the very steep cut from 30 Hz, and also demonstrates the slight resonant effect which is typical of this type of circuit. In this case, however, the resonance is very well-controlled, of the order of 1 dB only. The phase-shift graph, Figure 5.42, shows the existence of resonance as a sharp and very large phase reversal.

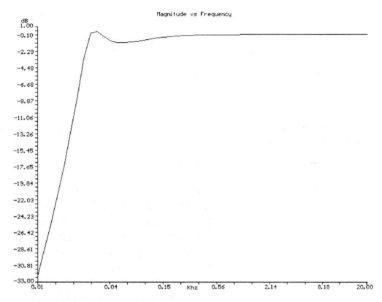

Figure 5.41 *The Aciran analysis of the Linsley Hood high-pass filter.*

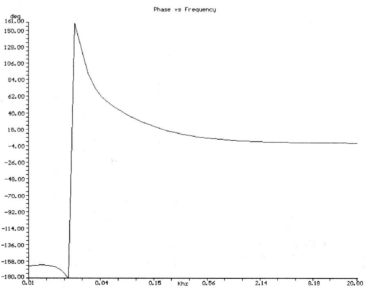

Figure 5.42 *The phase characteristic for the active high-pass filter.*

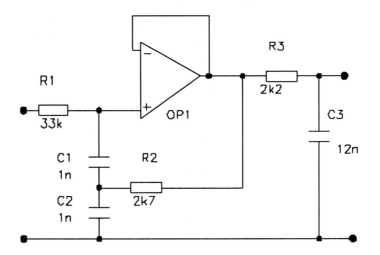

Figure 5.43 *The Linsley Hood low-pass active filter circuit.*

The low-pass filter of Figure 5.43 uses the same circuit methods to achieve a hiss filter with a turnover frequency of 10 kHz and the same slope of 20 dB per octave. The circuit is a mirror image of the previous one, with capacitors and resistors interchanged, so that much the same type of response might be expected.

The graph of amplitude plotted against frequency in Figure 5.44 shows that the slope starts gradually, with its –3 dB point at around 10 kHz, but with the resonant kink in the response also at around this

Figure 5.44 *The amplitude response plotted by Aciran for the low-pass filter.*

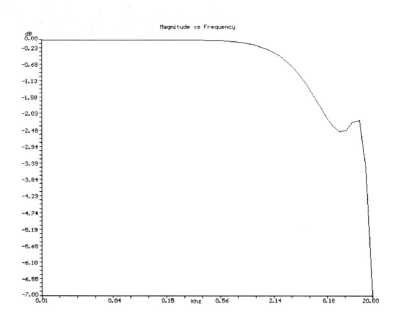

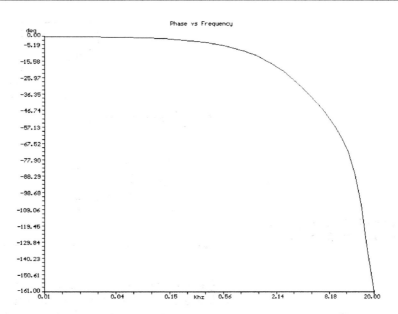

Figure 5.45 *The phase graph for the low-pass filter.*

frequency before the steep part of the slope. The phase graph in this example shows, curiously, a gradual slope increase with no sudden reversals (Figure 5.45). Once again, this demonstrates that simply rearranging the components of a circuit, interchanging resistor and capacitor positions, does not necessarily produce a mirror-image response. As before, however, the twitch in the amplitude response curve is insignificant in decibel terms.

6 Last lap

The examples provided along with Aciran have not so far been considered because they involve no effort on your part and are therefore not ideally suited to learning how to mark out nodes and enter component values. They are, however, excellent examples of how Aciran can tackle circuits that would otherwise present quite exceptional difficulties, and in this section, some of the examples will be used to show circuit responses that are not made clear in the brief documentation that accompanies Aciran.

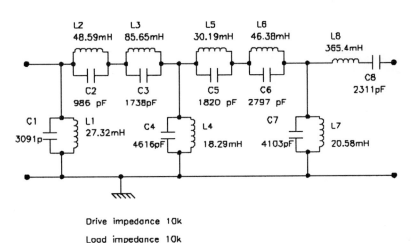

Figure 6.1 *The circuit for the Aciran EXAMPL5.CCT file, an elliptical filter circuit consisting entirely of passive components.*

Figure 6.1 is the circuit for the file EXAMPL5.CCT, an elliptic-function bandpass filter which uses a large number of tuned circuits to establish a bandpass characteristic. This is the type of circuit which calls out for analysis by computer, because anything else would be immensely difficult and time-consuming, even if formulae were available. There are eight tuned circuits in this example, all calling for calculated values which in practice probably would be substituted by nearest preferred 5% values. This would alter the response, making an Aciran analysis even more valuable.

The document file of Aciran suggests analysing this example with a linear response over the range 12 kHz to 24 kHz in 20 steps, but the analysis looks better when a larger number of steps is used, Figure 6.2. The bandpass characteristic is most impressive, with very steep sides, and though the top looks rather jagged, the variations are less than 1 dB and therefore insignificant. The behaviour is less smooth at the fringes, but since these are all between –56 dB and –80 dB with respect to the passband they also cause no problems in the overall response.

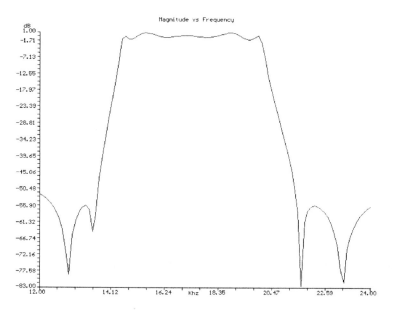

Figure 6.2 *The amplitude vs. frequency plot for the elliptical filter, made using 100 steps rather than the suggested 20.*

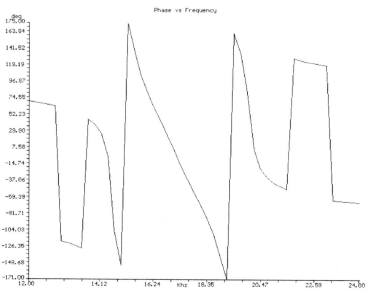

Figure 6.3 *The complicated phase response for the elliptical filter circuit.*

Figure 6.3 shows the phase response of this example. As you would expect, this shows a large number of sharp reversals, each due to one resonant circuit, and the input impedance magnitude graph, Figure 6.4, is also complex, with two prominent peaks at which the normal input impedance of around 10k rises to more than 50k. The output impedance graph, Figure 6.5, follows an opposite path, with the nominal 10k impedance rising to 27k at the extremes of the frequency range. An

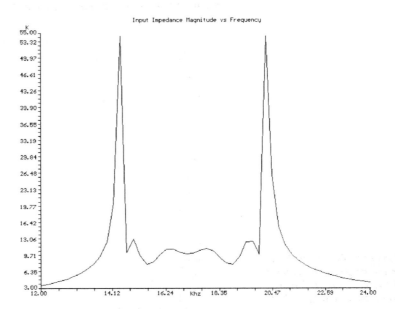

Figure 6.4 *The input impedance variations for the elliptical filter.*

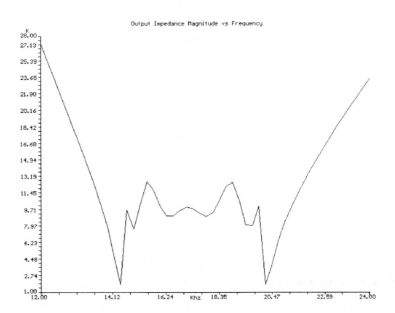

Figure 6.5 *Output impedance magnitude plotted for the elliptical filter.*

analysis of VSWR shows that a value of unity is achieved for both input and output in the passband, rising (as would be expected) to high values outside the passband.

Figure 6.6 shows the Aciran example of an active delay line. The aim of the circuit is not to provide a response whose amplitude varies with frequency, but one in which the real part of the amplitude is almost constant and the phase alters in proportion to frequency, so that time delay is almost constant for signals in the frequency band that is being used. In this sense, the circuit acts as a delay line would.

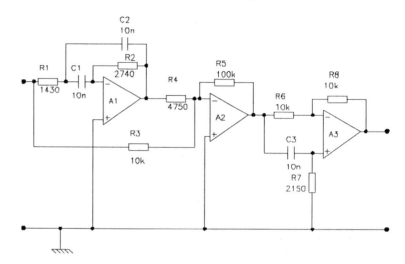

Figure 6.6 *The Aciran active delay line example.*

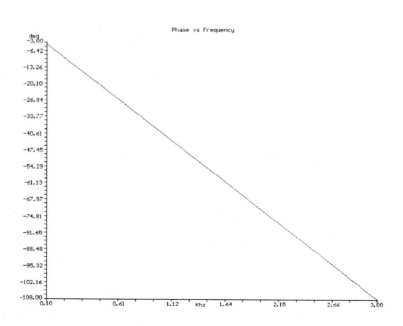

Figure 6.7 *The phase vs. frequency response for the active delay line, showing the almost ideal straight-line shape.*

The output magnitude is virtually flat at +20 dB for the whole of the frequency range of 100 Hz to 3 kHz. Figure 6.7 shows the phase response, almost a straight line when plotted against frequency on this linear scale. This is the requirement for equal delay, so that the graph of time delay also shows an almost steady value (Figure 6.8). The graph looks far from straight at the higher-frequency end, but this is due to the stretched vertical scale that has been used – the actual change is only 0.2 µs in a delay time of 100 µs.

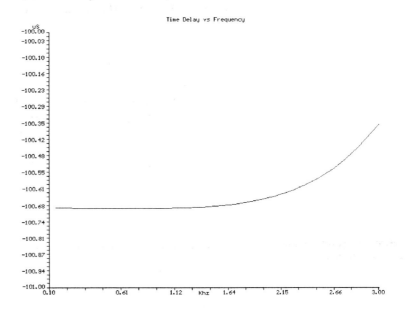

Figure 6.8 *The time-delay plot for the active delay circuit.*

The most complex of the Aciran examples is the 5-stage active low-pass filter whose circuit is difficult to follow when printed out from the documents on the disk. The diagram, drawn using AutoSketch, is shown in Figure 6.9, and is a good indication of how complex a circuit can be analysed. The circuit consists of five dual OPamp circuits, each self-contained and linked by resistors. As before, several of the resistor values are calculated values, and in practice it is likely that 1% standard

Figure 6.9 *The 5-stage active low-pass filter circuit example.*

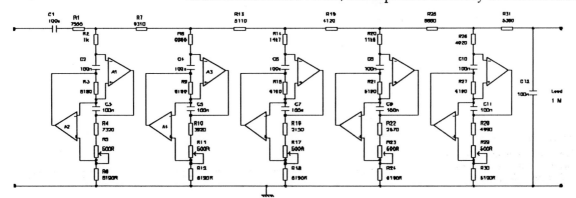

values would be used, and that the potentiometers would then be used to trim the circuit until its performance was at the optimum level. All values of resistance are in ohms unless otherwise indicated.

The complexity is emphasised during an analysis run from 100 to 300 Hz linear, because even on a fast AT machine it will take a minute using the recommended 20 steps, and on slower machines it can involve a minor coffee break. The amplitude graph, Figure 6.10, shows the steep cut that is obtained and the phase graph, Figure 6.11, indicates how many resonant circuit equivalents are being used.

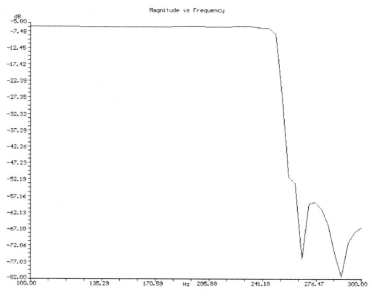

Figure 6.10 *The amplitude graph for the active low-pass filter, showing the excellent response shape.*

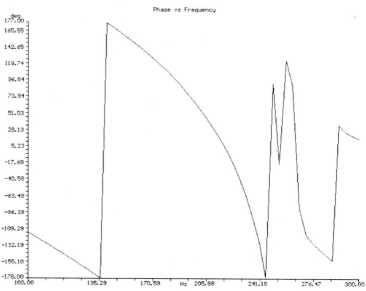

Figure 6.11 *The phase response for the active filter.*

Using resonant lines

The older versions of Aciran did not allow the use of resonant lines as circuit components, barring Aciran from work on many modern RF circuits and communications work. Version 3.0 has remedied this, with a line specified as a component for which three values are required in addition to the two nodes. These values are:

Z_0	The characteristic impedance for the line
L	The line length in centimetres
E_r	The relative permeability for the line material

For lines constructed using normal materials and in the absence of ferromagnetic materials, the value of E_r can be taken as unity. The factor is available for use with specialised lines for which the permeability is not the same as that of free space.

To work with the line as a component you need to know enough line theory to set up Aciran for analysis. Lines are used as tuning and matching components, particularly for signals in the UHF band where line lengths can be short. The figures for characteristic impedance can be obtained from manufacturers' handbooks – they are usually in the 50-300 ohm region. Working with lines will usually involve cutting lines to some fraction of the wavelength of the signal to be used, so that you need to be able to find this wavelength. This is given in centimetre units by:

$$\lambda = \frac{c}{f}$$

where λ is wavelength in centimetres, c is 3×10^4 and f is frequency in MHz. For example, the wavelength for a frequency of 1000 MHz is 30 cm. Lines cut for matching purposes may be used open circuit or short-circuited at the unconnected end, and the short-circuit version (length different from the open-circuit version) is always to be preferred because there is then no chance of the end radiating. The two examples included with Aciran illustrate the main types of circuits, and the examples below are of a similar type.

Figure 6.12 illustrates a section of line being used to match two other sections which have different characteristic impedances. When a line is terminated with its characteristic impedance at each end, energy

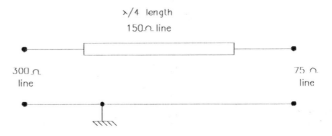

Figure 6.12 *A resonant quarter-wave line being used to match impedances.*

travelling along the line is not reflected, so that standing waves are not set up. This corresponds to a voltage standing wave ratio (VSWR) of unity, and the aim of most line designs is to achieve results while maintaining the VSWR as close to unity as possible. If lines of different impedances are directly connected, reflections will be set up where the lines join, and this will result in a VSWR greater than unity. The VSWR can be restored only by matching the different impedances, the line equivalent of using a transformer.

Line theory shows that to match two different impedances a section of line can be used whose length is equal to a quarter of the wavelength of the signal and whose impedance is the geometric mean of the impedances being matched. As an equation, the impedance required, Z_0, is given by:

$$Z_0 = \sqrt{Z_1 Z_2}$$

The example shows an input of 300 ohms impedance required to match with an output at 75 ohms for a frequency centred around 100 MHz. Since the wavelength for 100 MHz is 300 cm, a quarter wavelength is 75 cm. The characteristic impedance of this section must be the square root of 300×75, which works out at 150 ohms. The analysis should check results over a range of frequencies around the central 100 MHz, perhaps 95 MHz to 105 MHz, and should check in a linear sweep that input and output impedances are correct and that the VSWR is low over the range.

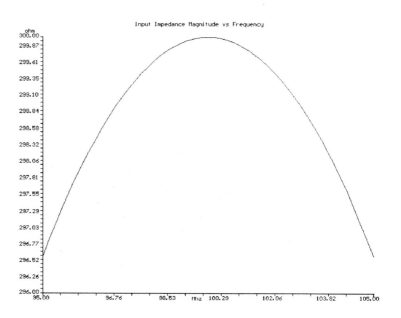

Figure 6.13 *The input impedance magnitude plot for the matching line section.*

Using the Aciran circuit description of this line circuit, the *Return loss* and *Impedances* options in the *Config* menu should be switched on, and the *Sweep* type set to *Linear*. Moving to analysis, try 25 runs from 95 MHz to 105 MHz. The input impedance graph is illustrated in Figure 6.13 above. This looks far from flat at a first glance, but the scales at the side show that there is little more than 3 ohms difference between the peak and the edges, so that the impedance matching is correct to 1%. An output impedance graph looks like an inverted version of the input graph, though centred around 75 ohms.

The input VSWR graph, Figure 6.14, shows a central value of unity, with the edge frequencies giving a value of 1.13, not a value that would be likely to cause problems. In practice, the irregularities at the cable connections would almost certainly raise the VSWR above this low value.

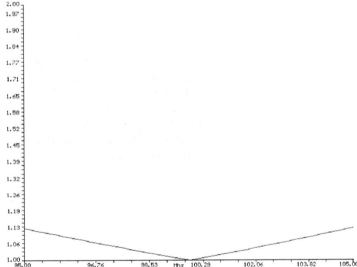

Figure 6.14 *The input VSWR plot for the matching section.*

Open and shorted lines

A line which is either open-circuit or short-circuit must result in reflections since it is not correctly terminated. These reflections mix with the normal forward wave to set up standing waves, meaning that permanent high and low voltage levels can be measured at fixed points along the line. This condition corresponds to a high VSWR figure (of more than 10) and though standing waves are not necessarily undesirable (in a tuned line) they make the line suitable for one narrow band of frequencies only (if at all).

When one end of a line is short-circuited, there can be no wave voltage at the short circuit, though the wave current will be a maximum. The standing wave therefore has a minimum voltage amplitude at this point, and at each half wavelength from that point. These places of zero voltage amplitude are called nodes. By contrast, if the end of a line is open-circuited, it can have a voltage maximum but no current, and there will be maximum voltage points, or antinodes, for each half-wavelength back along the line.

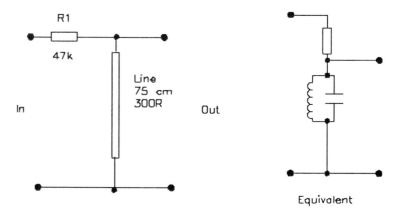

Figure 6.15 *A line and its equivalent resonant circuit.*

A resonant line behaves like a tuned circuit, and this can be demonstrated by the circuit of Figure 6.15, in which a piece of line is fed by a 47k resistor. The line is 75 cm long and shorted at one end, so that it should be one quarter of a wavelength long at 100 MHz. This should cause the open end of the line which is connected to the resistor to be equivalent to an open circuit at this frequency, so that the whole

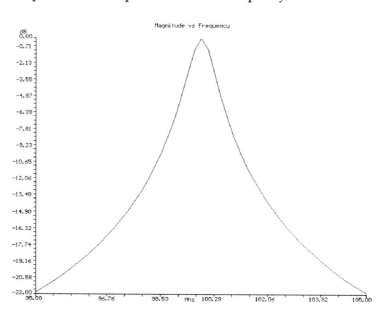

Figure 6.16 *The amplitude vs. frequency plot for the resonant line circuit.*

circuit should be like that of a parallel tuned circuit fed by a resistor, as indicated in the drawing.

When this is analysed by Aciran, specifying the line as 300 ohms and 75 cm long, the amplitude response graph of Figure 6.16 (above) looks very much as you would expect for the equivalent LC parallel circuit. The input VSWR plot, Figure 6.17, shows how sharply this quantity increases when a line is resonant.

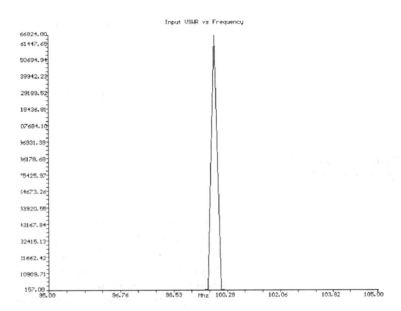

Figure 6.17 *The input VSWR plot for the line around resonance.*

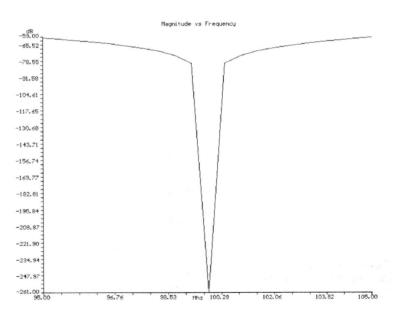

Figure 6.18 *The notch filter action of a line terminated with a high resistance value.*

The line can also be used as a notch filter, using the line with a load resistor, and yielding the amplitude response of Figure 6.18. The corresponding phase plot is illustrated in Figure 6.19.

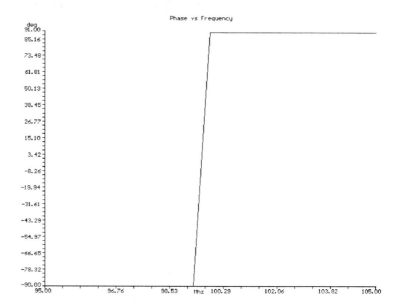

Figure 6.19 *The phase response of the line notch filter.*

Imperfect components

Aciran models some components in a fairly simple way, and the most unrealistic of these models is the transformer. The transformer model performs a step up or down of voltage which is proportional to the ratio of turns, and which is uniform for all frequencies. In order to make this realistic, inductance and capacitance has to be added to the model. This is not done by default because in some applications it can be useful to insert a perfect transformer as a substitute for other components as, for example, to apply phase-split inputs to a circuit.

A real transformer has a finite winding inductance for each winding, and the wire used for the windings also has resistance. In addition, some of the signal in the primary has no effect on the secondary, and this can be simulated by a leakage inductance. For tuned transformers it can be useful to show the inductance of both windings, but for untuned transformers the equivalent of Figure 6.20 is more useful. This regards all imperfections as being in the primary circuit, with winding inductance, leakage inductance and series winding resistance all shown in the primary. Coupled to this set of real components is an ideal transformer which simply carries out the voltage transformation action. The quantities used for the components in such a diagram have to be calculated from measured transformer parameters, and where a set of values are shown for one winding only, they are said to be reflected into that winding. In the example, leakage, resistance and winding inductances of the secondary are shown reflected into the primary, a process which involves multiplying each secondary quantity by the square of the turns ratio Primary:Secondary.

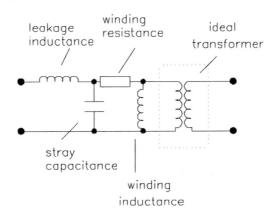

Figure 6.20 *An equivalent circuit for an untuned transformer.*

The equivalent circuit for the transformer can be simplified for special cases, because some factors are more important at low frequencies than at high frequencies and vice versa. At low frequencies, the stray inductance and stray capacitance can be ignored, and the winding inductance is the important factor, shunted by the reflected load resistance. At high frequencies, the winding inductance is not important unless it is likely to resonate with the stray capacitance, and the leakage inductance and stray capacitance have a considerable effect on the way that the transformer behaves.

The snag with all this, as with so many simulations, is finding the information. Manufacturers' manuals will give full information on a component such as a transformer, but the catalogues of distributors seldom do, unless the applications for the transformer are those which require exact specification. If you are winding your own transformers, you are very much on your own.

Figure 6.21 shows a transformer used to match signal into an 8 ohm load which might be a loudspeaker. At low frequencies we can ignore leakage inductance and stray capacitance and concentrate on winding

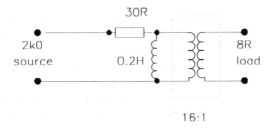

Figure 6.21 *An impedance-matching transformer, showing the low frequency equivalent.*

inductance and resistance. The source impedance is 2k0 in this example so that the ratio of turns is:

$$\sqrt{\frac{2000}{8}}$$

giving approximately 16:1 as the ratio of the (ideal) transformer. We can see how a primary (plus reflected secondary) inductance of 0.2H and a series resistance of 30R will affect the ideal situation of 100% transfer. When this is typed into Aciran, some care has to be taken over nodes and the winding ratio. The primary of the transformer can be earthed at one end of each winding so that it is Node 0. The winding ratio must be shown as Primary:Secondary, but with the primary as 1. A 16:1 ratio must therefore be entered as 1:0.0625, which is 1 to $\frac{1}{16}$. The analysis amplitude graph of Figure 6.22 is taken using a linear sweep, with 2k as the generator impedance and 8R as the load impedance, between 10Hz and 150Hz. The result is illustrated in Figure 6.22, showing a well-behaved frequency roll-off in the bass region.

When a transformer equivalent circuit for high audio frequencies is used the ratio can again be taken as 16:1, translating to 1:0.0625, and this time the primary inductance, the leakage inductance, resistance and stray capacitance are the important features. When this is analysed using 2k source resistance and 8R load resistance, the circuit also appears to be well-behaved with an almost flat response from 1 kHz to 30 kHz. This is because of the damping effect of the 8 ohm load, and in real life a loudspeaker load has inductance and capacitance as well as

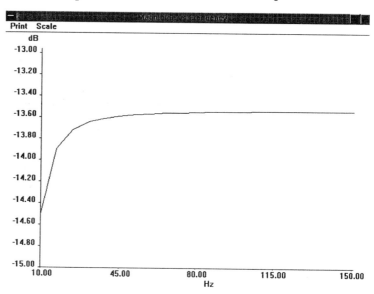

Figure 6.22 *The transformer circuit analysed for low-frequency response.*

resistance. Unless these quantities can be estimated with some accuracy and a realistic circuit equivalent made, Aciran cannot show the effects of using a transformer in a circuit of this type. A model of this type should take secondary inductance into account rather than reflecting its (low) value into the primary circuit.

Semiconductor simulations

The simple model of transistor that is used by Aciran is useful for a large amount of work that does not require operation near the frequency limits of the transistor. When this cannot be assumed, the transistor has to be modelled as an equivalent circuit, of which the current source or g_m equivalent is an essential part. The current source component can be shown in a diagram as a current generator with two nodes labelled as 'From' or 'Source' and 'To' or 'Sink', along with a positive and a negative control node.

In this form, a simple transistor amplifier can be drawn in the format shown in Figure 6.23, the circuit used as Aciran example number 5. The source point for the g_m unit is at the collector of the transistor and the sink is the emitter; the control + node is the base and the control − node is the emitter, making a three-terminal device, shown within a dotted outline in the drawing.

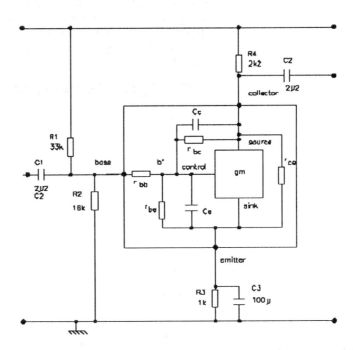

Figure 6.23 *The small signal equivalent Hybrid-Pi model of a transistor.*

Using this type of equivalent depends on finding values for the various equivalent parameters which are:

Rbb	Base spreading resistance
Rbe	Base-emitter resistance
Cbe	Base-emitter capacitance
Rbc	Base-collector resistance
Cbc	Base-collector capacitance
Rce	Collector-emitter resistance

Some of these values are very large, particularly the base-collector resistance, usually 10-100M, which provides some negative feedback in circuits of high impedance, and this quantity is often ignored in circuits which use low input impedance values. The series base spreading resistance, of 100 ohms or so, is considered negligible when the input impedance levels are of several k. The collector-emitter resistance is usually ignored unless unusually high output impedance values are being used. The capacitances are not so negligible, however, particularly the base-collector capacitance which, at around 10 pF, limits the high frequency response of the transistor through Miller feedback effect – in the conventional model this is the cause of the f_T value. The Cbe value, though around 50 pF, is often negligible when the input impedances are low. The capacitance values become very significant when the transistor is being operated at high frequencies, and it is at such frequencies that this model becomes particularly useful.

By way of comparison, Figure 6.24 shows the Aciran example of a straightforward single transistor amplifier analysed in both ways, with

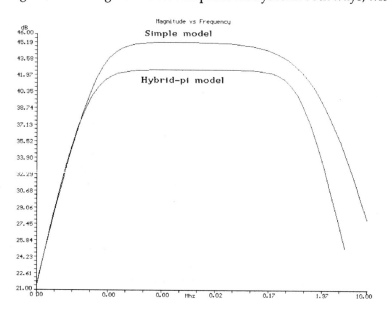

Figure 6.24 *The amplitude frequency plot for a simple single transistor amplifier, comparing standard and Hybrid-Pi models.*

the graphs superimposed as far as possible. The use of the Hybrid-Pi model produces less gain, a factor of around 3 dB in this example, and though the low frequency end of the analysis is almost identical for both models, the Hybrid-Pi model shows a more rapid fall-off at the higher end of the bandwidth. The phase-shift graph, Figure 6.25, displays much greater differences between the respective models, with the simple model showing a smooth phase change to an ultimate value of −180° and the Hybrid-Pi model showing a violent change of phase.

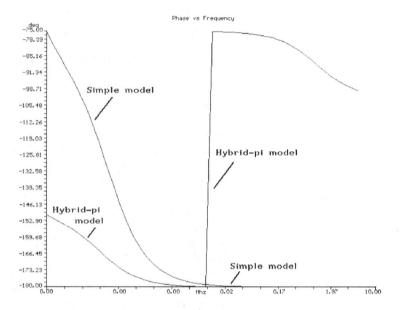

Figure 6.25 *The phase plot for the two transistor models shows more interesting discrepancies.*

Aciran registered version

When you register with the author, James Herron, you will receive the most recent update of Aciran and in addition another version of the Aciran program, known as TAC. This is Tiny or Turbo Aciran which runs some four times faster than the normal version. TAC is fully compatible with the main version and can use the same files, but it is limited to fewer nodes. This allows the program to work within one 64 Kb block of memory which, on a PC machine, ensures much faster operation. Unless your requirements are for large circuits (which is unusual for linear circuits since any circuit can be broken down into smaller portions) you should use TAC so as to save time, particularly if you want to carry out multiple analysis runs to take tolerances into account.

Another way of speeding up the work of Aciran is to use a suitable co-processor in your computer. If your computer has a co-processor it will be detected automatically by Aciran and used to perform all calculations. No changes need to be made to the program itself to make use of a co-processor, but you have to ensure that your computer can make use of the co-processor chip and that you use the correct type of chip. At one time, this was simple and straightforward, but now that several manufacturers offer a proliferation of co-processor chips selection is rather more difficult.

If you fit a co-processor after installing Aciran you will have to re-install Aciran to gain the benefits of the co-processor with Aciran.

It is important to emphasise that a co-processor is not essential, and its cost is of the order of £40 to £200 or more, depending on the type of chip that is used. A co-processor is desirable if your requirements are intensive, which is likely if you use Aciran, a CAD program like AutoSketch or AutoCAD and any type of spreadsheet, all for professional purposes. Hobby users of Aciran do not need a co-processor unless they are determined to have the fastest linear analysis in their street.

The type of co-processor that is used depends on the machine. The older type of XT machine which uses the Intel 8086 or 8088 processor can make use of the Intel 8087 co-processor, and the two main items to consider are whether there is a socket for the chip on the main board of the computer, and the speed of the main processor. Most computer main boards nowadays are fitted with sockets for a co-processor of the correct type, but older machine may not possess such a luxury, in which case you really need to upgrade the whole machine; it is not really feasible to install a co-processor socket as an add-on. The original clock rate of the IBM PC/XT was around 4.77 MHz, and any variety of 8087 co-processor should be able to work with such a clock rate, but later machines have used considerably faster clocks, and you may need to buy a co-processor suitable for the clock rate of your computer. There are complications involved in this for Amstrad owners, because the early Amstrad machines (PC1512 and PC1640) used an 8 MHz clock rate for the main processor, but ran other chips at half this speed. Take advice from the supplier of the co-processor chip before you buy.

The situation is even more complicated for owners of AT machines which use the 80286 or 80386 chips. The corresponding co-processors are the 80287 and 80387, but the 80386 chip can make use of either the 80287 or the 80387 co-processor, and you need to know what provisions are made on the main board. In addition, the range of clock speeds used in AT machines is very large, ranging from as low as 8 MHz on some early models to 33 MHz or more on modern 386 machines. Once again, you need to check with the supplier of the co-processor that the chip you want to fit will be suitable. Several manufacturers, notably Intel, IIT and Cyrix, supply co-processor chips which are not of identical design (though they use the same pin-outs), and you will need to make a decision for yourself on price, availability and performance.

The ACTRAN conversion utility

In this book, all input to Aciran has made use of the keyboard. ACTRAN is a conversion utility which allows you to input ORCAD files to Aciran. ORCAD is a large and costly circuit design package, not available as shareware, which is capable of providing an output of circuit information in a number of CAD formats. The ORCAD output file, called a *netlist* file, can then be used by other CAD systems such as a PCB layout package. All the information needed by Aciran can be obtained from an ORCAD netlist in a format known as SPICE; see the notes at the end of this chapter. If you are a professional circuit designer and have used ORCAD to produce your circuit diagrams, it is particularly easy to use Actran to convert the netlist file into an Aciran circuit file, rather than to enter the whole circuit again by hand.

The ORCAD system comes complete with a number of component libraries, and these may not all be in the correct format for SPICE netlist files. Normally these would need modification, but you can make use of an Aciran library called ACIRAN.LIB, which contains most of the elements used by Aciran in the SPICE format. For more detailed information on ORCAD library formats you should consult your ORCAD handbook for the version you are using.

Using ACTRAN

Ensure that ACIRAN.LIB is placed at the start of the list of library files used by the Draft portion of ORCAD. This will ensure that Draft will look in ACIRAN.LIB first for any components that it uses. Create your circuit using ORCAD in the normal way. Since Aciran circuits have one input and one output node it is necessary to inform Aciran what these nodes are in the ORCAD circuit. This is done by calling one input module port 'input' and one output module port 'output'. The node numbers corresponding to these modules are placed in a MAP file by the Netlist utility of ORCAD so that they can be read by ACTRAN.

Most of the circuit examples supplied with Aciran are given in their ORCAD form of *.SCH file as well as the Aciran *.CIR circuit files, but the procedure for working with a new circuit, assuming that you have just produced a circuit using ORCAD, is as follows:

1. You must annotate and carry out an electrical rules check on your circuit, and carry on only if no errors are found.

2. Use the Netlist utility to generate a netlist file in the SPICE format, by typing:

   ```
   Netlist <CircuitFileName> <NetlistFileName> /S Spice
   ```

   Press the ENTER key to carry out the action. The Netlist utility supplied with ORCAD will then produce two files, a netlist file

(*.NET) and a map file (*.MAP), both of which are used by ACTRAN.

3. Once these files are on the disk you are ready to run ACTRAN. At the MS-DOS prompt type ACTRAN and press the ENTER key. The ACTRAN screen will appear and you will be prompted for the netlist filename. You can use a full path name, such as C:\ACIRAN\NETFILE\CIR1.NET. If ACTRAN finds the file it will report on its progress on the output window while it processes the netlist information.

If any unknown component is found by ACTRAN it will be ignored and a warning will be issued. ACTRAN also produces a LOG file, which contains a record of all the screen output, so that you can check back on any warnings. Once ACTRAN has completed this work you should examine ACTRAN.LOG to check for any errors, using the TYPE command in the form:

```
TYPE ACTRAN.LOG | MORE
```

This allows you to look at a page at a time. Alternatively, if you are using MS-DOS 5.0 (or later) you can use the DOSSHELL utility and select View File Contents to see the ACTRAN.LOG file. This, unlike TYPE, allows you to scroll up and down and examine any part of the file at leisure.

There is the possibility that you will be using types of Transistors, FETs and OPamps that are not listed in the Models directory. ORCAD uses only the designator of Q for both Transistors and FETs, so that ACTRAN cannot distinguish one from the other. Since bipolar transistors are now being used to a lesser extent, ACTRAN assumes that such a Q-component is a FET and will check in the Models directory for a FET of the given type. If a FET of that type is not found, ACTRAN then tries to find a transistor of the given type. If the component type is still not found ACTRAN leaves it as a transistor and gives it the name STANDARD, using the model file STANDARD which contains information on a general-purpose NPN transistor. ACTRAN tells the user that this substitution has taken place.

A similar situation occurs when ACTRAN is unable to find model files for OPamps. All OPamps have the designator U?, and ACTRAN only expects to find OPamps ICs and therefore the presence of other integrated circuits will only lead to confusion. After completing the transfer, ACTRAN will ask for a circuit Description of up to 30 characters. This is simply the Name that you normally add to an Aciran circuit file to identify it.

Next you will be asked for a filename and you should enter any valid Aciran circuit filename. ACTRAN will now store the circuit file in your Circuits directory. It is important how the values of some components are entered in the ORCAD sheet so that ACTRAN can read them correctly. The majority of components are straightforward, but Transformers and Transmission Lines need more attention, and you should study the example files provided in the ORCAD directory.

Note that Transformers require the turns ratio to be supplied as:

```
<primary turns> <space> : <space> <secondary
turns> ^ colon
```

Transmission lines require their parameters to be stated as:

```
Zo=<char impedance> <space> L=<length in cm>
<space> Er=<permeability>
```

The colon and equals signs are important, as well as the order of parameters.

Notes on PSPICE

PSPICE is a circuit analysis program of vast size and capability, and a version that is sold as Public Domain software is for demonstration purposes only, limited to a circuit of ten transistors. The importance of PSPICE is that other programs, including Aciran, can make use of its file structures, and that for the serious professional user with $1000 to spare, it is the standard system for analysis, linear or non-linear. PSPICE and its companion printing program, PROBE, were devised by the Department of Electrical Engineering and Computer Sciences in the University of California at Berkeley and made commercially available. The following is a very brief account of this extremely large and complex program.

SPICE is a general-purpose circuit simulation program, which can handle non-linear DC (bias), non-linear transient (pulses), and linear AC analyses. The circuits that are analysed may contain resistors, capacitors, inductors, transformers (or coupled inductors of any type), independent voltage and current sources, four types of dependent voltage or current sources, transmission lines, and the usual semiconductor diodes, bipolar transistors, junction FETs and MOSFETs. There are built-in models for the semiconductor devices, and the user need specify only the pertinent model parameter values – the Public Domain version provides very few, however.

A more advanced model of the junction transistor is used, reverting to the familiar Ebers-Moll type if there are insufficient parameters specified. In either case, charge storage effects, ohmic resistances, and a current-dependent output conductance may be included. The diode model is suitable for either junction diodes or Schottky barrier diodes. The junction FET model is the type described by Shichman and Hodges. Three MOSFET models are implemented, one using a square-law I-V characteristic, one an analytical model, the third based on experimental data, with the latter types including second-order effects such as channel length modulation, sub-threshold conduction, scattering limited velocity saturation, small-size effects and charge-controlled capacitances.

The DC analysis portion of PSPICE determines the DC bias operating point of the circuit with inductors considered as being shorted and capacitors treated as open circuits. A DC analysis is automatically performed before any other analysis. This is done before a transient analysis to determine the transient initial conditions, and before the AC small-signal analysis to determine the linearised, small-signal models for non-linear devices. The DC small-signal value of transfer function, input resistance and output resistance can also be computed as a part of the DC solution if required. The DC analysis can also be used to generate DC transfer curves by specifying an independent voltage or current source which can alter value in steps over a specified range, with the DC output variables stored for each of the source values. In addition, PSPICE can also calculate the values of DC small-signal sensitivity for output variables with respect to circuit parameters. These DC analysis options can be specified on lines in the file that start with .DC, .TF, .OP and .SENS. In general, each action of PSPICE is specified by a line in a file that starts with a dot followed by a command code.

By using a line starting with .OP it is possible to see the small-signal models for non-linear devices in conjunction with a transient analysis operating point. The DC bias conditions will be identical for each case, but the more comprehensive operating point information is not available to be printed when transient initial conditions are computed in this way.

The AC small-signal portion of PSPICE computes the AC output variables as a function of frequency. The program starts by calculating the DC operating point for the circuit and then determines linearised, small-signal models for all of the non-linear (active) devices in the circuit. The resultant linear circuit is then analysed over whatever range of frequencies you specify. The result of an AC small-signal analysis is usually required to be a transfer function such as voltage gain or trans-impedance. If the circuit has only one AC input, it is convenient to set that input to unity amplitude and zero phase, so that output variables have the same value as the transfer function of the output variable with respect to the input.

PSPICE can also simulate the generation of white noise by resistors and semiconductor devices, using the AC small-signal portion of PSPICE. Equivalent noise source values are determined automatically from the small-signal operating point of the circuit, and the contribution of each noise source is added at a given summing point. The total output noise level and the equivalent input noise level are calculated for each frequency point you have specified. The output and input noise levels are normalised with respect to the square root of the noise bandwidth and have the units Volts/Hz or Amps/Hz. The output noise and equivalent input noise can be printed or plotted like any other output variables, and no additional input data are necessary for this analysis. Flicker noise sources can be simulated in the noise analysis by including values for the parameters KF and AF on the appropriate device model lines.

A particularly valuable option for linear circuit designers is the ability to compute the distortion characteristics of a circuit in the small-signal mode as a part of the AC small-signal analysis. The analysis is performed assuming that one or two signal frequencies are imposed at the input, with the frequency range and the noise and distortion analysis parameters specified on the .AC, .NOISE and .DISTO control lines.

The transient analysis portion of PSPICE calculates the transient output variables plotted against time over a specified time interval, with the initial conditions automatically determined by a DC analysis as usual. All sources that are not time-dependent, such as power supplies, are set to their DC values. For large-signal sinusoidal simulations, a Fourier analysis of the output waveform can be specified to obtain the Fourier coefficients as a function of frequency. The transient time interval and the Fourier analysis options are specified on the .TRAN and .FOURIER control lines.

All input data for PSPICE is assumed to have been measured at 27°C (300 K) and the simulation also assumes this same nominal temperature. The circuit can be simulated at other temperatures by using a .TEMP control line. This option is very useful because of the considerable changes in semiconductor parameters with temperature.

The input format for SPICE is of the free format type, with each line of instructions divided into fields that are separated by one or more blanks, a comma, an equal (=) sign, or a left or right bracket sign; extra spaces are ignored. A line may be continued by entering a + (plus) in column 1 of the following line so that PSPICE will continue reading that line starting with the second column.

A field containing a name must begin with a capital letter (A to Z) and must not contain any of the dividing characters (delimiters) mentioned above. Only the first eight characters of the name are used, so that there is no point in using longer names, though they will not cause any problems.

A field that contains a number may be an integer field (such as 9 or –6), a floating-point field (such as 6.7714), either an integer or floating-point number followed by an integer exponent (1E-8 or 1.44E4), or either an integer or a floating-point number followed by one of the scale factors: T, G, MEG K, M ,U, N, P or F as used also in Aciran. Letters that immediately follow a number and which are not scale factors are ignored, and letters immediately following a scale factor are also ignored.

The circuit to be analysed is described to PSPICE by a set of element lines that define the circuit layout and the component values, and a set of control lines that define the model parameters and the run controls. The first line in the input set must be a title line, and the last line must be a .END line, but the order of the remaining lines is not important except where a line is being continued.

Each component in the circuit is specified by a component line that contains the component name, the circuit nodes to which the component is connected, and the values of the parameters that determine the electrical characteristics of the component. The first letter of the component name specifies the component type. For example, a resistor name must begin with the letter R and can contain from one to eight characters. Hence, R, R1, RSE, ROUT and R3AC2ZY are valid resistor names, allowing you to use the same nomenclature in PSPICE as on the circuit diagram.

Data fields that are enclosed in angle bracket signs < > are optional. Where punctuation marks such as brackets, equal signs, etc. are indicated in a PSPICE example, these must be correctly used. Where branch voltages and currents are used, PSPICE always uses the convention that current flows in the direction of voltage drop.

A node is marked by an integer (whole) positive number, but the numbers need not be in sequence. The earth node must be numbered zero as in Aciran. The circuit must not contain a loop of voltage sources and/or inductors, and each node in the circuit must have a DC path to earth. Each node must have at least two connections except for transmission line nodes (so as to permit the use of unterminated transmission lines) and MOSFET substrate nodes.

The Title Line, such as RF MOSFET CIRCUIT, must be the first line in the input set. Its contents are printed exactly as typed as the heading for each section of the output. The last line must be .END, with the dot in the first screen column. You are allowed to place comment lines, which start with an asterisk (such as * AIM FOR GAIN OF 80) in any part of a file.

Component Lines take the format of name, nodes, value, options. For example, the line R5 6 3 220 means that resistor R5 is connected between nodes 6 and 3 and has a value of 220 ohms. One or two temperature coefficient figures can follow the resistor value if temperature dependence is to be calculated. Capacitors and inductors follow a similar pattern but with no temperature coefficient allowance. The value of an inductor or capacitor can, however, be followed by an initial current or voltage value if this is required (only if the UIC option is specified on a .TRAN line). Non-linear capacitors and inductors can be described if the function for inductance or capacitance is known, describing how inductance or capacitance varies with current or voltage respectively.

Transformers use a line such as TX2 L3 L4 0.82 which shows the two (or more) inductors that are coupled and the coefficient of coupling between them. The inductors are separately described with a dot placed on the first node of each inductor. Transmission lines are treated as being lossless and require name, nodes, characteristic impedance, and a quantity that expresses the length of the line, using time delay or the combination of frequency and electrical length at that frequency. Two lines can be used to simulate the case where two modes of a line are excited. Optional initial conditions can be specified to provide the voltage and current at each of the transmission line ports, but this option is used only if the UIC option is specified on the .TRAN line.

PSPICE also allows circuits to contain linear-dependent sources which use transconductance, voltage gain, current gain or transresistance.

Semiconductor devices are most easily dealt with, as they are in Aciran, by using a separate .MODEL line and a unique model name for each different type. The device component lines in PSPICE then refer to the model name. This avoids the need to specify all of the model parameters on each device component line. Each semiconductor device component line then contains the device name, the nodes to which the device is connected, and the device model name. In addition, other optional parameters of geometry and initial condition can be specified for each device, particularly area factor for diodes and channel, drain and source diffusions for MOSFETs. Initial conditions can be used when circuits contain more than one stable state or when transient analysis is required.

A very useful aspect of PSPICE is that you can specify the use of a sub-circuit (of PSPICE components) much as the semiconductor model file is used. The sub-circuit is defined in the input file by a grouping of component lines and the program then automatically inserts that group of components wherever the sub-circuit is required. There is no limit on the size or complexity of sub-circuits, and sub-circuits may contain other sub-circuits. A sub-circuit definition is begun with a .SUBCKT line.

The file will also contain control lines, such as a .TEMP line. For example, a line such as >TEMP -10 20 100 will specify the temperatures at which the circuit is to be simulated, using temperatures in Celsius. A .WIDTH line specifies the output print width. An .OPTIONS line allows the user to reset program control and user options for specific simulation purposes.

The .OP line will force PSPICE to determine the DC operating point of the circuit. This is in addition to the automatic analysis that is performed prior to a transient analysis to determine the transient initial conditions, and prior to an AC small-signal analysis to determine the linearised, small-signal models for non-linear devices, so that PSPICE will perform a DC operating point analysis if no other analyses are requested. There are other command lines for DC transfer function and other PSPICE actions.

As an example of a PSPICE file, the following, taken from the PSPICE manual is used to determine the DC operating point and small-signal transfer function of a simple differential pair. In addition, the AC small-signal response is computed over the frequency range 1 Hz to 100 MHz.

```
SIMPLE DIFFERENTIAL PAIR
VCC 7 0 12
VEE 8 0 -12
VIN 1 0 AC 1
RS1 1 2 1K
RS2 6 0 1K
Q1 3 2 4 MOD1
Q2 5 6 4 MOD1
RC1 7 3 10K
RC2 7 5 10K
RE 4 8 10K
.MODEL MOD1 NPN BF=50 VAF=50 IS=1.E-12 RB=100 CJC=.5PF TF=.6NS
.TF V(5) VIN
.AC DEC 10 1 100MEG
.PLOT AC VM(5) VP(5)
.PRINT AC VM(5) VP(5)
.END
```

7 Starting with AutoSketch

Requirements

AutoSketch 3.0 for MS-DOS or AutoSketch for Windows both enable you to produce drawings of professional quality on paper as large as your printer or plotter can handle – for most desktop printers this will be A4, but pen-plotters can be bought in much larger sizes at correspondingly higher prices. Your drawings can include text such as labels, headings and the symbols of mathematics, music and other specialised applications. All of this can be done in the office or at home using any PC-compatible computer.

☐ AutoSketch 3.0 requires considerably less storage space on the hard disk, and this can often be a significant factor in deciding which version to use if your computer normally works using Windows. AutoSketch 3.0 for DOS can be run from the Windows system if desired.

You also need a video screen system capable of displaying graphics. AutoSketch can be used with any of the standard systems such as CGA, EGA, VGA or Hercules, but the systems which offer higher resolution, such as Hercules and VGA, are much easier to work with, and VGA is virtually standardised on modern PC machines. You must have a mouse connected to the computer and with suitable (driver) software installed. You also need the use of a printer or plotter. Work of good quality can be produced using a dot-matrix printer, but much better results can be produced using a laser printer, inkjet printer, or a pen-plotter such as the Hewlett-Packard, Roland or the remarkably inexpensive ACS-APT type.

For large drawings, it is an advantage on the older type of PC machines (using 8086 or 80286 microprocessors) to fit an expanded (not *extended*) memory board to your computer. AutoSketch 3 can use up to 2 Mb of such memory. For fast drawing actions, it is an advantage to fit a maths co-processor chip. Modern machines of the 80386 or 80486 type can use large amounts of memory that can be configured either as extended (the default) or expanded. For use with AutoSketch 3 for MS-DOS, extra memory should be configured as expanded, but if your machine is used at all times with the Windows system it might make more sense to configure all of the extra memory as extended and use AutoSketch for Windows (see Appendix B).

The use of a hard disk is essential. It is *just* possible to work in a crippled way with twin 1.2 Mb or 1.44 Mb drives, but for any serious purposes the use of a hard disk of at least 40 Mb (and preferably more) is virtually essential. Your drawings can be saved in ordinary format or in a special DXF format which can be interchanged with AutoCAD. Text can be interchanged with a word processor, and drawings can be exported to a desktop publishing (DTP) program.

☐ For any machine running Windows, a memory of 5 Mb and hard disk of 150 Mb or more is required to obtain the full benefit from the system. A machine using only MS-DOS can make use of 1 Mb of memory and a smaller hard disk.

Drawings made using AutoSketch can show as much detail as your *printer* can deal with. Ordinary screen-painting programs can only show as much detail as can be seen on the screen, limited to about 72 dots per inch. Using AutoSketch with a laser or inkjet printer allows you to use up to 300 dots per inch, and with a plotter even finer detail can be achieved.

AutoSketch is particularly suited to *scale* drawings in which each centimetre on the drawing (on paper) must correspond to some dimension (one foot, one metre, ten metres, etc.) of the real-life object that is being drawn. This makes AutoSketch ideal for room or garden plans, engineering drawings, plans for yachts, furniture, models, clothes or whatever you like. It can also be used for maps, and for items which need no scale, like holiday charts, planning diagrams, electrical circuit diagrams and so on. In this book, the descriptions of AutoSketch are devoted exclusively to its use for circuit diagrams, so that features which are of interest in mechanical or architectural drawing (like showing scale distances) are not mentioned, or mentioned only in passing.

☐ All the circuit diagrams in this book were originated using AutoSketch 3.0 running under MS-DOS.

Installation

Before you can start using AutoSketch 3 you need to install the program on your hard disk. The program files are supplied on either 5¼" or 3½" disks, and the number of disks depends on the type of disks.

☐ The older type of 5¼" 360 Kb disks are virtually obsolete, so that unless you are using an old machine you should ensure that your copy of AutoSketch comes on 5¼" high-density, or 3½" disks. Many modern machines are equipped only with the 3½" type of drive which can read either low or high-density 3½" disks.

You should make backup copies of these disks in case of damage to the originals. The process described is applicable only if your computer can use the same size and type of disks as the AutoSketch distribution disks.

☐ If your computer uses a 5¼" HD drive, it can read the older type of 5¼" low-density disks, but it cannot copy them.

To back up, using compatible disks, proceed as follows:

1. Ensure that your computer is using the C: drive (displaying C>) and that you have a set of blank disks of the correct type at hand (formatted or unformatted).

2. Place the first AutoSketch disk into drive A:.

3. Type DISKCOPY A: A:, and follow the instructions about inserting the destination disk (the copy) and the source disk (the AutoSketch disk).

4. Repeat this copying action for all of the disks in the set.

AutoSketch 3 automatically creates a directory when you install it on a hard disk. Do *not* try to copy the AutoSketch files onto your hard disk, because they are held in coded form and are decoded during installation.

1. Place the first AutoSketch disk into the A: drive and type INSTALL (press ENTER).

2. You will be asked about disk directory names – accept the default of SKETCH3 by pressing the ENTER key.

3. You will be asked to confirm whether or not your computer uses a maths co-processor chip (the INSTALL program senses such a chip, but you need to confirm). Do not claim to have a co-processor unless you are certain that it has been installed – PC computers as sold are not fitted with this chip, and it will be present only if you (or a previous owner) have fitted it.

4. Remove each disk in turn and insert the next disk as requested.

5   Confirm that you accept the changes to the AUTOEXEC.BAT file, and to the creation of a SKETCH3.BAT file at the end of the process.

☐ If you did *not* opt, during installation, to have the AUTOEXEC.BAT file of the computer altered, you may find that the batch file that AutoSketch 3 creates does not work, so that typing SKETCH3 (ENTER) does not start the program. If this happens, either add ;C:\SKETCH to the PATH line in the AUTOEXEC.BAT file or alter the SKETCH3.BAT file so that it appears as:

```
set asketch=C:\SKETCH3\SUPPORT
set asketchcfg=C:\SKETCH3
C:\SKETCH
SKETCH %1 %2 %3
CD C:\
```

The AUTOEXEC.BAT file is a text file that is located in the C:\ (root) directory of the computer. It can be altered by using any program described as an Editor, and the altered version will be used next time the machine is switched on or reset (by pressing the Alt-Ctrl-Del keys). Do *not* use a word processor to alter AUTOEXEC.BAT unless the word processor is set to save files in the form called ASCII or DOS. See any good book on MS-DOS, such as the *Newnes MS-DOS Pocket Book* (Butterworth-Heinemann) if you need help with these alterations.

Configuration of AutoSketch 3

AutoSketch 3 must be configured (adjusted) before use for the type of mouse, type of screen and type of plotter or printer that you have fitted to your computer.

☐ Configuration of AutoSketch for Windows is not needed because the Windows system has already been configured with the necessary information.

Configuration is done simply by selecting numbers from each of several lists. Start the program, with the C> sign showing, by typing SKETCH3 (if the PATH line in the AUTOEXEC.BAT file has been edited to include the SKETCH directory) or by using CD \SKETCH (press ENTER) followed by SKETCH (press ENTER). You will then see the options under the following headings.

Pointer

Select a pointing device. For a mouse, select *Microsoft Mouse* because most other brands of mouse use the same system as the Microsoft mouse.

CONFIGURATION OF AUTOSKETCH 3

☐ Use any other number only if you are certain it applies to the system you are using. The use of keyboard keys (no mouse) is not really satisfactory.

Display

Select the type of *Graphics Card* your computer uses – check with the manual for your computer. Most modern machines use VGA.

☐ Note that the menu displays the full names of these devices, but your manual is likely to use the initial letters only: for example, VGA for Video Graphics Array.

You can then select the colour scheme for a colour display. For a monochrome display (which is clearer to see) the only options are normal (black lines on white background) or inverse (white lines on black background). Inverse can be kinder on the eyes because the screen is less glaring. For colour displays, you will have several options of foreground and background colours. Reply Y to the *Activate scrollbars* question if you are using a mouse.

Printer or plotter

Select the printer or plotter you intend to use.

1. If you are using a dot-matrix printer and your make of printer is not shown in the list, use the Epson option. Laser printers can usually be set either to emulate Hewlett-Packard LaserJet or Post-Script standards. If you are using a plotter, it is most likely to emulate the Hewlett-Packard plotter.

2. Select the most suitable option, and you will be asked to specify a model in more detail, particularly for dot-matrix printers. There are also several options for the LaserJet models.

3. Specify the connection method, usually parallel. If you are forced to use serial connection, you will have to fill in details of the settings used for the printer or plotter, taken from its manual.

4. Select the *name* of the printer/plotter connection. Use LPT1 for a parallel connection and COM1 for a serial connection, unless you know that a different method is necessary (perhaps LPT2 if there are two parallel printer connectors, or COM2 if there are two serial connectors).

☐ If at any time in the future you need to alter these settings, do so by starting AutoSketch as follows:

    Type CD \SKETCH3; press ENTER
    Type SKETCH /R; press ENTER

The alternative is to find the file called SKETCH.CFG and to delete it.

Using the mouse in AutoSketch

☐ This section assumes that your machine is fitted with a mouse and the mouse has software (such as MOUSE.COM) correctly installed.

The mouse will have either two or three buttons on its top surface, depending on the make of mouse. Only one button need be used, normally the left-hand button. As the mouse is moved on the desk or mouse mat the pointer on the screen will move so that it can be placed on words in the menu bar. The word will appear in inverse (colours reversed) to show that you are pointing to it. The pointer in AutoSketch will be either an arrow or a finger, depending on the type of action last used.

Tap the left-hand button (press and release it quickly). This is *clicking on the item*, and its effect is to activate the item that has been selected. Clicking on the *Assist* word in the menu bar will cause the *Assist* menu to appear on the screen. This is a comparatively small menu of six items only (Figure 7.1). Some of the names on the list are accompanied by a letter/number combination, such as 'A6'. This means that, taking this example, pressing the Alt and F6 keys together will have the same effect as selecting with the mouse. 'F6' by itself would mean the F6 key alone.

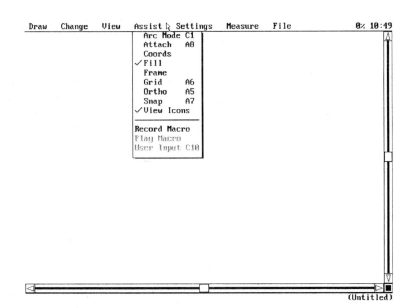

Figure 7.1 *The starting screen appearance of AutoSketch 3, showing the Assist menu clicked.*

☐ To remove this or any other menu from the screen, move the mouse so as to place the pointer on a blank piece of screen and then click.

A tick mark is used in a menu when an action can be switched on or off. The tick shows that the action has been switched on, and if this item is selected again, the tick will be removed, showing that the action is off. In the illustration the actions called *Fill* and *View Icons* are shown ticked.

☐ It is particularly important to have the *View Icons* item ticked, since this makes it easy to locate drawings in the directories.

Some menu items appear in grey or dotted print, and are described as being 'greyed out'. This means that these choices are not available at present. When you are using the options that these items refer to they will be printed in solid colour.

Dragging the mouse

Dragging is a mouse action that is done unusually in AutoSketch, not in the way that is familiar to users of the Windows system. The scroll bars are at the right-hand side and the bottom of the screen.

Click on the box in a scroll bar. This will make that box follow the movement of the mouse (along the bar). The box is released when you click again (you do *not* need to keep the button depressed while you move the mouse).

This *dragging* action will move the drawing on the screen, either up/down or left/right. Some menus of AutoSketch also allow you to see more items by dragging the box of a vertical scroll bar.

You can also click on the pointer at either end of a scroll bar, or on the space between the box of the scroll bar and either end of the bar. This will move the drawing on the screen, and the extent of the movement depends on the method. Clicking on an arrowhead will move the drawing by one quarter of the screen width or height (depending on whether you are using a horizontal or a vertical scroll bar). Clicking on the space between the box and the end of the bar will move the drawing by half of the screen width or height.

Limits

AutoSketch deals with dimensions as numbers when you make a drawing, with no reference to units (feet, inches, metres, millimetres) until you come to print or plot the drawing. The *Settings Limits* box allows you to specify the largest range of dimensions on the normal drawing screen, but this does not restrict you to remaining within these limits.

☐ The importance of *Limits* is for setting the number of drawing units. Dimensions in AutoSketch use these drawing units right up to the time when the drawing is printed.

For circuit diagram work, assuming that you are working on A4 paper (using a printer) the *Limits* should be set up for this size as follows:

1   Click on the *Settings* menu. This brings down a long list of options (Figure 7.2).

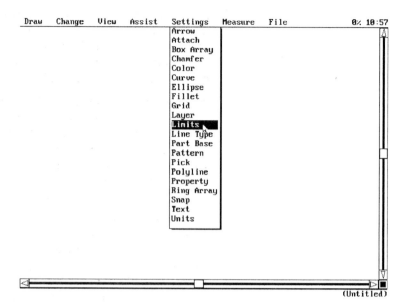

Figure 7.2 *The options of the Settings menu.*

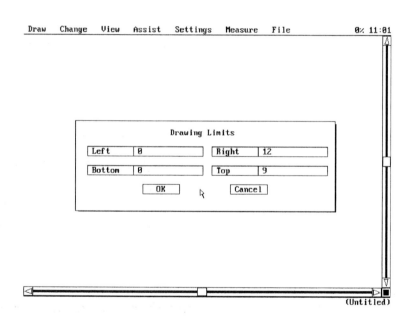

Figure 7.3 *The Drawing Limits box as it first appears.*

2   From this set, click on *Limits*. This bring up the *Drawing Limits* box, Figure 7.3. The settings are shown as 12 x 9 for use with inch measurements. Alter *Right* to 200 and *Top* to 150 to suit A4 paper size in millimetres for the first set of examples.

☐ Note that these are smaller than A4 dimensions to allow a drawing to be made up to 200 mm wide and 150 mm deep, with the paper the normal way up (*Portrait*, as distinct from *Landscape*). These settings do not restrict the drawing size, they only establish the normal limits of drawing units.

☐ It will be an advantage for the examples in later chapters to use larger limits, such as 400 x 300. This will be useful if you want to be able to draw thick lines: see Chapter 9. Do not be tempted to use very large numbers as limits because this will, later, cause problems with *Grid* and *Snaps* settings. Using large numbers as limits makes the unit of drawing smaller, and this can make it more difficult to carry out actions that depend on locating a point exactly.

The alteration of *Limits* is carried out by clicking to the right of the number and using the backspace key to delete the figure. Type the new figure and click on OK in the same line to confirm (Figure 7.4). You can opt to cancel if you want to. When both figures have been altered, click on the OK box at the bottom of the form. The illustration shows the right limit of 200 typed into place, waiting for you to confirm by clicking on OK.

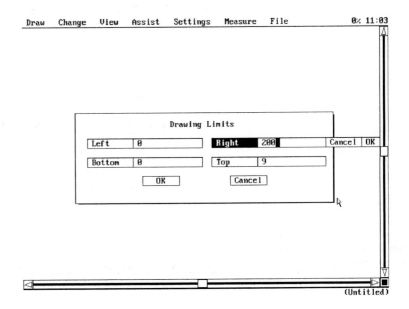

Figure 7.4 *Editing the numbers in the Drawing Limits box.*

Measuring units

The *Units* options, also in the *Settings* menu, allow you to specify whether you want to work in feet/inches or in decimal units. For *any* decimal units – which can be feet, inches, meters or any other units – select *Decimal* and specify the number of digits of decimals following the point. This is usually set by default to 3. For circuit diagram work you can specify zero as the number of decimal places.

☐ You do not need to specify the units as mm with *Decimal Suffix* unless you want to print units as well as dimensions on diagrams. This is not needed for circuit diagrams, so that the *Decimal Suffix* box can be ignored.

The drawing screen

The screen that appears when you run AutoSketch (after configuring) is the Drawing Screen, which contains the Menu Bar, scroll bars, and the arrow pointer. All of your drawing work is done on this screen display, but there are several options about the information that is seen on the screen.

1. Lines will meet each other exactly only if *Attach* has been selected, or some other form of snaps (see later) used. Do not assume that you can manipulate the mouse with sufficient accuracy to ensure that you can join lines precisely.

2. The grid of fine dots on the screen can be turned on or off using the *Assist* menu, and the spacing of the dots is determined from the *Settings* menu. If you have set *Limits* as described above, click on *Settings* and *Grid*, and set the grid spacings to 4 and 'On' as shown in Figure 7.5. When you set the *X Spacing* and click on the OK to the side of it, the *Y Spacing* will be set to the same figure so that it does not need to be altered.

☐ Too close a spacing cannot be set and you will be informed if you have selected too close a grid spacing, such as 1, and the grid display will be turned off. Remember that when the drawing is printed you will be using 1 mm as your unit for printing purposes. If you use drawing limits of 400 x 300, a suitable grid setting is 8.

☐ The display of co-ordinates at the foot of the screen is in terms of your drawing units, and is also switched on or off from the *Coords* item in the Assist menu. Co-ordinates are distances measured in drawing units from the bottom left-hand corner of the screen. The X co-ordinate is the distance right (+) or left (–); the Y co-ordinate is the distance up (+) or down (–). The *Coords* figure is always shown with four places of decimals even if you have opted to use fewer places in the *Units* menu.

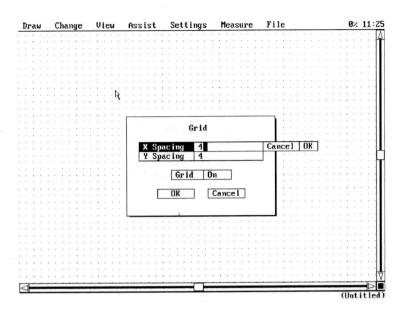

Figure 7.5 *Setting the grid spacing to 4 and turning display on.*

The Attach option

This can be switched on or off by selecting *Attach* from the *Assist* menu. This allows you to place either end of a line precisely, something that is not possible simply by moving the mouse. When *Attach* is switched on, you can draw a line to either end, or at the exact centre of an existing line, or the intersection of lines. A circle can have a line, or any other shape, attached at its centre or at any of four points around the rim.

All shapes (see later) have their own attach points to which another line end can be precisely located. Use *Frame* from the *Assist* menu to show the attach points for a curved shape.

☐ *Attach* is not necessarily beneficial for circuit diagrams because you very often want to draw lines to or from points other than the ends or middle of another line, for example. Use *Attach* only when you know where the Attach points are and need to use them. You can switch *Attach* on during the course of a drawing, even with a line half-drawn. For most purposes you will be using *Snaps* instead of *Attach*.

Snaps

The *Snaps* option is an important one, because it forces the cursor to move only to pre-determined points, such as each drawing unit. When *Snaps* are selected on from the *Assist* menu, the arrow cursor of Auto-Sketch 3 is replaced by a cross marker. This marker can be moved only to snap points, which are by default the drawing unit points. For example, if your drawing is 200 mm x 150 mm, there will be 200 snap points across and 150 snap points down, a total of 200 x 150 = 30,000 snap points. The *Snaps* form follows the same pattern as the *Grid* form, and does not need to be separately illustrated.

If you are working, as you ought to, with *Grid* on, the default grid size will be the same as the snap size, so that your cursor will snap to each grid point. When you alter the grid size, however, this does not alter the *Snaps* size. A separate menu action is needed to alter the *Snaps* size, and for circuit drawings it can be very useful to have different sizes, with a *Snaps* setting of 1 rather than the setting of 4 which is used for *Grid*. For *Limits* of 400 x 300, you could set *Grid* to 8 and *Snaps* to 2. Too small a *Snaps* setting will make it difficult to ensure that you are locating a point correctly, because even a very small movement of the mouse will cause the pointer to snap to the next position.

☐ You do not have to conform to these suggested settings, but they have proved particularly suited to circuit diagram work.

For some purposes it may be better to turn *Snaps* off. This is particularly true when a circle is being drawn, such as the enclosing circle for a transistor or FET, because using *Snaps* on forces the radius of the circle to be a multiple of the snap unit. For all other purposes, however, it is better to keep *Snaps* switched on, and this includes locating the centre of a circle.

☐ When *Snaps* are being used and the display of co-ordinates is on, the co-ordinates numbers will be simpler. For example, if you are working with 200 mm x 150 mm limits and have *Snaps* set to 1 (meaning 1 mm in this case), then the co-ordinates numbers will take values such as 79.0000, 10.0000, 130.0000 and so on, rather than the fractions with four significant places of decimals that you see when *Snaps* are not being used.

Always try to work with *Snaps* on as far as possible, however, because this ensures precision which cannot easily be obtained otherwise. It is also an advantage to have *Coords* switched on along with *Snaps*. Using *Coords* switched on ensures that you can place the pointer at the same X or Y settings when you move it to another part of the screen, so allowing you to line up parts of a drawing correctly.

Using AutoSketch

If you have never used any form of Computer Assisted Drawing, the most daunting part of the exercise is starting to draw anything. Drawing circuit diagrams has the considerable advantage that you will be working with a set of standard symbols that are familiar to you, but you are left with the problems of creating these symbols. Though symbol sets can be bought, they are more likely to be to U.S. drawing standards, and suited to inch units, than sets you create for yourself.

In the following chapter, then, we will tackle the drawing of basic shapes, showing how these actions can be used to form electronic component symbols. It is very important to keep to the same units for drawing all symbols, so that the sizes are consistent, and this is aided by the fact that when a drawing is saved as a disk file, the *Limits*, *Grid* and other settings are saved with it and will all be recovered when the file is loaded again.

All drawing methods in AutoSketch 3 follow a similar pattern.

1. The pointer or cursor is used to select the shape that is to be drawn.

2. One point, such as a line end or the centre of a circle, is located by moving the mouse and then clicking to establish the point.

3. The mouse is moved (dragged) to another point which will be the end of a line or the rim of a circle, and the mouse button is clicked again to establish this point.

4. The same set of point and click actions can then be used to draw another shape of the same type without the need to re-select. A drawing action remains in use until it is cancelled by another command.

5. Any drawn shape (object) can be undrawn, erased selectively, moved or copied, along with a number of other shape- and size-changing actions.

8 Drawing and changing shapes

Preparation

Before you start making a drawing, go through the following essential steps:

1. Set up your limits, using a number of drawing units that will be convenient.

2. Set up the grid size and snap size.

3. Click on the *View* menu, and then on *Zoom Limits*. This constructs the drawing outline that should be ideally suited for circuit diagrams.

☐ IMPORTANT: If you then select the *File* menu, then *Save As*, you can fill in a filename such as TEMPLATE so that this group of settings can be restored whenever you want them. If you do not do this, your settings will have to be keyed in each time you start AutoSketch.

Using a template like this, with nothing drawn on it, can be a very convenient method of enforcing uniformity of settings. If you later want to change the template you can load the template file in, alter the settings and then save it again. Remember, however, that any drawings made using the template will save their settings with the drawing, so that you will see the older template settings when you load in an older drawing made with these settings.

Drawing shapes

All drawing of shapes is started from the *Draw* menu – once you have opted for a type of shape, the same shape can be repeatedly drawn without the need to re-select until you make another menu choice. Always use *Grid* on and *Snaps* on unless you particularly need to work without them, usually for drawing small circles. The *Point* option allows you to place a dot wherever you click the mouse button, but the dot will be very small and for a laser printer set to its 300 dots-per-inch resolution and a single dot made in this way is practically invisible when printed. See later for details of creating larger dots for use in circuit diagrams.

Lines

To draw a straight line, select *Line* and position the cursor to where you want the line to start. Click the mouse button and move the cursor to where you want the line to end. The line will stretch and change angle (*rubber-banding*) as you move the mouse. Click again to fix the endpoint.

☐ Click again on the same point if you want this to be the start of another line – remember that there is no need to select the *Line* option again if you want to draw another line.

The ends of the line will always snap to your snap points if *Snaps* has been turned on. This makes it particularly easy to draw shapes using straight lines (Figure 8.1). The example shows capacitors, earth symbols, and an OPamp symbol. You could also add diodes, thyristors and Triacs. It is an advantage to have each symbol drawn in two versions, one with a horizontal axis and the other with a vertical axis. This does not require you to draw each symbol twice, because AutoSketch allows you to make a perfect copy which can then be rotated (see later).

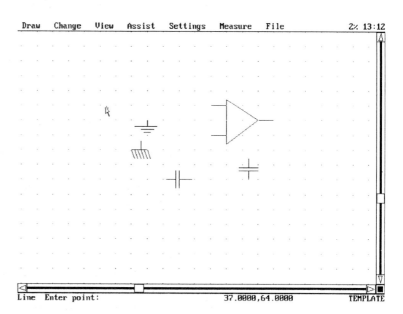

Figure 8.1 *The Line drawing command in use, and some of the typical component shapes that can be drawn. This is a magnified version using a Zoom (see later).*

☐ Be guided by the co-ordinate numbers when you want to make lines of equal size, such as the capacitor plates. Very often, the vertical lines look longer on screen than the horizontal, and the same applies to spaces. This effect is due to the limitations of the resolution of the monitor, and the dimensions will be shown correctly on paper.

At the moment, simply draw the component shapes without too much regard to being exact. Later we shall show how a complete set of component shapes can be drawn so that they will all be in scale and ready for making circuit diagrams. This requires the use of some techniques that we are not yet ready to look at.

☐ One point to note is that a shape constructed from lines is not considered by AutoSketch as a single shape – any of the lines can be deleted or moved out of place. Later we shall see how a set of lines (or other shapes) can be formed into a single shape, a *Group*, that cannot be broken into pieces.

The Polyline

A simple, or zero-width, *polyline* is drawn like a set of lines, ending – when the mouse is clicked – on the starting point again. The important difference between this and an ordinary set of lines is that a polyline counts as a single shape, not as six separate shapes. The polyline is more appropriate than the simple line for drawing the triangular outline for a linear IC. There is another version of Polyline, the wide Polyline, which is used for drawing thick lines, which can be single straight lines if required. See Chapter 9 for details of this action.

Box

To draw a box shape, as for a general IC symbol, select *Box* from the *Draw* menu and click on the position where one corner of the box will appear. Move the mouse so that the cursor is placed where you want to have the opposite corner and then click to create the shape. The box will always have its sides vertical and horizontal, so that if you want diamond shapes you can either draw four straight lines, or a polyline, or rotate a box (see later). The main application for the box shape in linear circuit drawing is to draw the base symbol for a transistor or substrate for a MOSFET, and for such symbols you may want to use a filled box (see later).

Circle

To draw a circle, select *Circle* from the *Draw* menu and click on the centre-point of the circle. Moving the mouse away from this centre point will draw a polygon (a set of straight edges). When the edge is where you want it, click again, and the polygon will change to a circle.

☐ If this is done with *Snaps* on, the rim of the circle will not necessarily be positioned exactly where you want it. The main use for the circle is as the casing for a transistor or FET. In such a case, always draw the straight line and box portions first, ensuring that the connectors will be on snap points, and then draw the circle later. Another option is to draw the circle quite separately and move it into place (see later). Figure 8.2 shows box and circle shapes added to the last view.

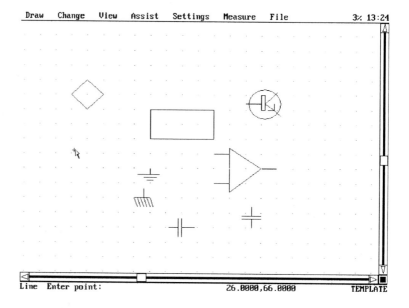

Figure 8.2 *Box and Circle shapes added to the previous set. Drawing these along with the other shapes helps keep dimensions compatible, essential if you are building up a set of component shapes.*

Arc, Curve and Ellipse

These shapes are not in much demand for drawing circuit diagrams, but it is useful to know how they are drawn in AutoSketch.

To draw an arc (part circle), select *Arc* from the *Draw* menu and click on the starting point, then on some point on the arc, and finally on the end point. The shapes that you can get are constrained by the *Snaps* settings. If you want more freedom about arc size, turn *Snaps* off *after* you have clicked on the first point of the arc.

For a *Curve*, click on a set of points on the curve, drawing a jagged line. This line will be smoothed into a curve when you click twice on the last point. With some practice, wave shapes can be drawn in this way – only a portion of wave need, in fact, be drawn with the rest obtained by mirroring and copying (see later).

For an *Ellipse* click on the centre point and then on one end of each axis (the longest or *major* axis and the shorter or *minor* axis). In the example, one axis is horizontal and one vertical, but these directions need not be used. Ellipses are seldom needed for drawing electronic diagrams. You may need to turn off *Snaps* after starting the ellipse if the use of *Snaps* makes the ellipse of the wrong size.

Figure 8.3 shows arc, curve and ellipse shapes added to the set.

☐ The arc can be used to generate inductor symbols: see later.

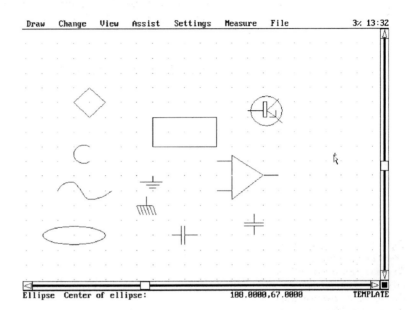

Figure 8.3 *The drawing with Arc, Curve and Ellipse added. Note that the top right-hand side of the screen shows the percentage of memory used and the current time.*

Undoing and breaking lines

It is often necessary to undo a drawing action, perhaps because the mouse has been moved too far. This is done from the *Change* menu, on which *Undo* is the first item. When *Undo* is clicked, the last performed drawing action will be deleted. Pressing *Undo* again will delete the drawing action before that, allowing you to delete all of a drawing in stages if you want. If you go too far along this path, selecting *Redo* will reverse the process for one drawing step.

☐ *Undo* and *Redo* are valuable, but their use requires storing the drawing steps on the disk. If a drawing requires a lot of disk space, it is possible to gain more usable space by turning off the *Undo* feature. This, however, has to be done when AutoSketch is started by using, in the batch file, the line:

```
SET ASUNDO=OFF
```

□ If you find that a drawing is taking an inordinate amount of disk space, save the drawing, and turn off *Undo*, by editing the batch file before you re-start AutoSketch. Load in the drawing and continue.

Another *Change* option is to break lines, particularly when careless use of the mouse has taken a line too far in one direction. AutoSketch allows this to be done in a simple way, using *Break* from the *Change* menu:

1. Select *Break*. A hand shape will appear, and the text at the bottom of the screen invites you to select the object. Place the pointer on the line which is to be broken (Figure 8.4) and click the mouse button.

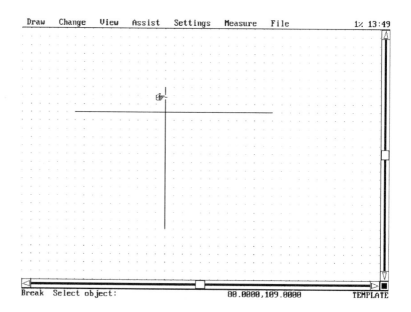

Figure 8.4 *The hand pointer that appears to allow you to point to the object: in this example, a line that is to be broken. Click to select.*

2. The line appears dotted. You are now asked to select the first breakpoint, Figure 8.5. Select, in this example, the point where the lines cross and click with the pointer there.

3. You are now asked to place the pointer on the other breakpoint and click again. This will break the line (Figure 8.6).

□ The selection using the pointer is greatly aided by using *Snaps* on – if the pointer is not touching the line, the nearest snap point on the line is used.

With *Snaps* on, a line can be broken only at a *Snap* position. If a line needs to be broken elsewhere – as, for example, where it crosses the rim of a circle – switch off *Snaps* temporarily while the breaking is being

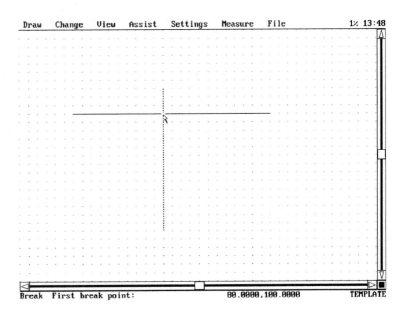

Figure 8.5 *The first breakpoint has to be selected; this can be either of the two ends of the gap.*

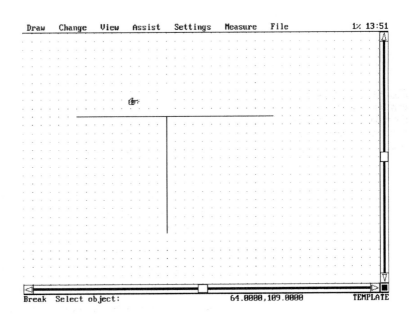

Figure 8.6 *The line broken. In the example, this makes a T-junction out of a crossing. Note that the bottom of the screen shows the start of another break action, because AutoSketch is always ready to repeat an action until another action is selected.*

done. Working with *Snaps* off is easier if you *Zoom* to the largest magnification that you can use: see later.

The way that *Break* works differs on closed shapes like circles and boxes. When you have chosen both breakpoints, the line that runs *anticlockwise* from the first break to the second will be broken. If you break the wrong section, use *Undo* to reverse the action.

☐ Lines and circles can be broken, but grouped objects, curves, text and objects that are filled cannot be broken.

Groups and parts

A drawing of an object, whether it is a full drawing or a detail of a full drawing, will normally consist of a set of lines and other shapes. These are considered as separate items and can be separately erased, copied, moved, etc. For example, a simple capacitor drawing consists of four lines, a resistor consists of a box and two lines, a transistor consists of a box, five lines and a circle, and so on. When these electronic component symbol shapes are drawn, each portion can be altered or moved, making the shape impermanent.

Such items can be grouped, making each into a single entity which AutoSketch treats as it would treat one line, selected as a single item. When a shape has been grouped, individual portions cannot be singled out for erasure, breaking or other changes, unless the whole object is first ungrouped, ensuring that actions on the shape do not distort the shape. Both *Group* and *Ungroup* are options in the *Changes* menu.

☐ Even fairly trivial objects in a drawing should be grouped because this ensures that they will always be treated as complete objects and there is no risk of accidentally selecting a single part of the object by mistake.

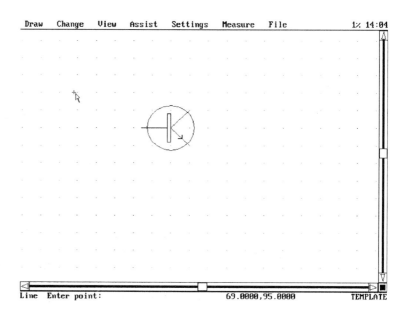

Figure 8.7 *A transistor shape as first drawn, consisting of lines, box and circle. This needs to be grouped so that it is treated as one object.*

Grouping is particularly important if a symbol has been drawn with a large number of small lines or other shapes. The more individual portions are used, the more important it is to group. Figure 8.7 shows a transistor shape (magnified view) which consists of a set of lines, a box and a circle. This was originally drawn as a box and five lines, but when the circle was drawn the line to the box (the base of the transistor) was too short, so that another joining line has been added. To group this:

1. Select the *Change* menu, and click on *Group*. You are asked to select the *Object*. This selection action is very important, as it is used extensively in AutoSketch.

2. The hand pointer is placed to the top left of the object and the mouse button is clicked. The mouse is moved, as if drawing a box, so that a frame appears around the object, Figure 8.8. The message: `Group Crosses/window corner` appears at the bottom of the screen.

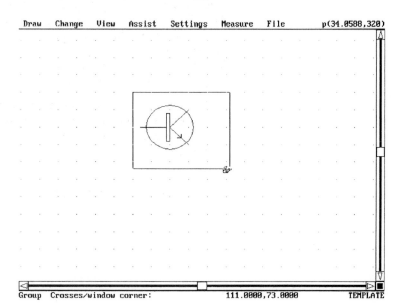

Figure 8.8 *The selection frame drawn around the object whose lines are being grouped.*

3. Click the mouse button to make the group. This makes the whole group look dotted, but the appearance will return to normal when another command is selected.

☐ Some care needs to be taken if you have more than one object – for example, a transistor symbol and a capacitor symbol – and you want to group each. It is tempting to group one, and then, with *Group* still selected, group the other. This, however, will make the transistor and the capacitor symbols belong to one single group, so that they cannot be separated. The correct method is to group one symbol and then click on the *Group* command again before grouping the second. You will see the normal appearance of the first symbol reappear when you click on *Group*, indicating that grouping is completed for that item.

In this example, a set of items has been selected by drawing a frame, and this is an alternative to selection by pointing to an object. When a shape consists of individual lines, any line can be selected for deletion or moving by using the pointer, but using a selection frame, as illustrated here, will ensured that a collection of objects (all those within the frame) is selected. Once a group has been formed, it can be selected with a pointer and it is no longer necessary to use a frame.

☐ The instructions above very explicitly required you to start the frame at the (top) left and end at the (bottom) right. This type of frame is called a *Window* selection box, and it selects only objects that are *wholly within* the box. Any line that is only partly within the box will not be included. If the box is drawn from right to left, it is a *Crosses* selection box, and it will include any object that the box outline crosses, as well as any object within the box. You are reminded of this by the words *Crosses/Window* which appear when you are selecting in this way.

Parts

An object, usually one that has been grouped, may be needed in many drawings. You will want to make electronic circuit diagrams, for example, using a standard set of symbols, each of which is a group of lines and other shapes, and you will not want to have to create these shapes from scratch each time you draw a circuit.

Any object which has been drawn – any object in a competed drawing, or even a complete drawing itself – can be saved as a *part*. This allows the drawing description to be saved to the hard disk, so that the shape can be recovered at any time later. When this is done, some point of the object must be designated as the *Part Base*, using the *Settings* menu. This forms a point for manipulating the object. The *Part Base* for a complete drawing is the bottom left-hand corner unless you select otherwise.

The object can be drawn alone, with nothing else on the screen, and saved in the usual way (so that there is no distinction between a drawing and a part, other than the use of a *Part Base*), or it can be a portion of a larger drawing and saved using *Part Clip*.

To use *Part Clip*:

1 Click on *Part Clip* from the *File* menu. In the form that appears (Figure 8.9), type a filename for the part. Names such as *Rhor* (horizontal resistor) or *Cver* (vertical capacitor) are useful reminders.

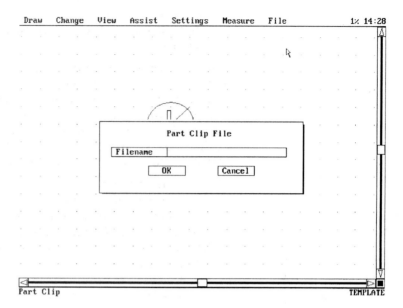

Figure 8.9 *The Part Clip filename form. Use a filename that reminds you of the type of symbol you are saving.*

2. You are asked to point to the part insertion base, Figure 8.10. When you load the part from the disk, this point will be at the pointer position.

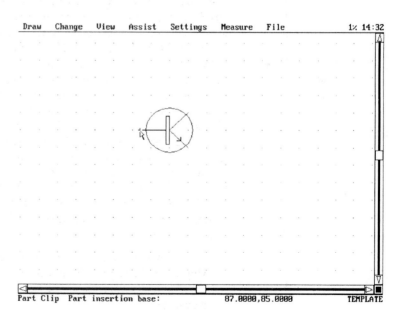

Figure 8.10 *Specifying the Part Base, in this example the end of the base lead of the transistor symbol, by pointing the arrow and clicking.*

210 DRAWING AND CHANGING SHAPES

☐ You need not be too fussy about placing the pointer exactly on the position, though the use of *Snaps* will usually ensure that the point is exactly placed. If you are working with *Snaps* off, get as close as you can and the *Pick* settings will ensure that the positioning is correct. The default *Pick* setting – see Chapter 9 – is that if the pointer is closer than 1% of screen height to a point, that point will be selected.

3. You are then asked to select the part, either by pointing to a grouped object or (as illustrated in Figure 8.11) by using a Window frame as before.

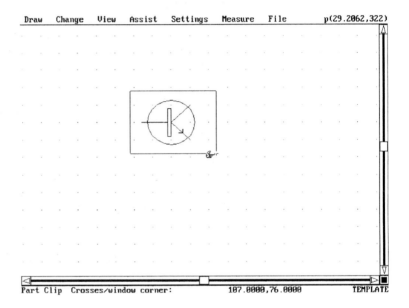

Figure 8.11 *Selecting the part with a Window frame.*

The object will be saved as a part, available to be placed into any future drawing.

☐ Though this sounds ideal as a way of preparing component symbols for drawings, it is a slow way to work, because each component has to be imported from the disk before it can be placed in a drawing. It is much better to make a drawing consisting of the component symbols that you need most urgently (resistors, capacitors, semi-conductors, inductors, switches etc.) and save the whole drawing. This can then be used as the basis of any circuit diagram. Only symbols that are in less frequent use need to be saved as parts, and these should always be drawn initially on the same screen as the other symbols to ensure that their dimensions and snap points are compatible.

When a part is needed, the *Part* option from the *Draw* menu is selected. This gives a list of all files, with icons (images), and the icon for the part can be selected. A pointing finger will appear, and the Part Base will be located at the end of the finger, so that this can be moved to any part of a new drawing.

☐ It is possible to buy disks of ready-made part files for popular symbols such as Electrical or Architectural. If you make your own parts you will need to ensure that they are created in a size that will be useful, and all to the same scale.

Creating a set of symbols

As hinted above, the best way of approaching circuit diagram creation is to make a drawing that consists of a set of the symbols, each made into a group, that are most commonly used. This drawing should be made with the desired *Limits* (such as 400 x 300 units), *Snaps* and *Grid* settings established, and it will then serve also to re-establish these settings each time it is loaded. A lot of time can be saved if each component is drawn in two positions (horizontally and vertically). In addition, this makes it easier to ensure that all of the components are drawn in compatible sizes.

Figure 8.12 shows a screen view of a set of symbols drawn using the methods that have been described so far. Each has been made into a group to avoid accidental alteration, but only after careful inspection. This inspection requires the use of *Zooms*.

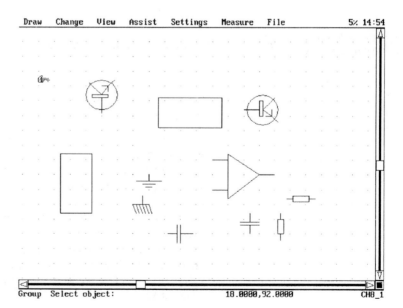

Figure 8.12 *A set of component symbols. Others can be added to this set later.*

Zooms

Zooming in AutoSketch means using the whole screen for a view of a drawing, or part of a drawing, at a different scale. This can be used to make the detail of part of a drawing more precise, or to make a small drawing fill the screen, or to make a drawing which is larger than the original limits fit into the screen space.

Because of the way that a CAD program is designed, zooming never makes lines look coarser, as it would on a conventional computer painting program. When you draw a line using AutoSketch, what is stored in the memory of the computer is an equation that describes the line. Magnifying a line simply involves multiplying the terms in the equation by a constant factor. In a painting program, each shape is described in terms of which dots on the screen are coloured, and when you try to magnify this, it is done by making the dots larger (not by making more of them) so that the drawing looks coarser. Zooming in AutoSketch therefore allows you to make fine adjustments to a shape, or to create symbols that are too small to be easily created at normal scale, such as dots.

☐ If you use *Zoom* to alter drawing scale in this way, you can return to the previous normal view by clicking on *View* and then *Last View*. You should always return to normal following a *Zoom*, because if you *Zoom* from a view that is already zoomed you will find it more difficult to return to the normal scale.

There are four *Zoom* options in the *View* menu: *Zoom Box*, *Zoom Full*, *Zoom Limits* and *Zoom X*. You can use the *Zoom X* option to alter the scale of the whole drawing by a set (number) factor. Numbers up to 1.0 will reduce the size of the drawing and numbers greater than 1.0 will increase the scale. The advantage of using *Zoom X* is that you *know* the size of the zoom factor – other *Zoom* methods will not necessarily use a whole number.

Zoom Box

This option, a very useful one for circuit diagram uses, allows you to select a portion of the diagram, as large or as small as you like, and expand it to fill the screen. Using the set of symbols as an example, the method is:

1 Select this option and use the mouse to draw a *Zoom Box* (Figure 8.13).

2 When you click on the second corner of the box the whole screen will be used to show the area of the box, presenting a magnified view (Figure 8.14).

You can alter *Grid* and/or *Snaps* (or turn both off) in order to make detailed alterations to this part of the drawing, but it is advisable to

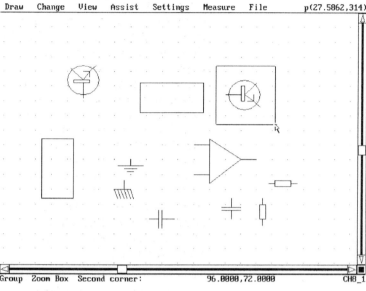

Figure 8.13 *A Zoom Box drawn around a symbol.*

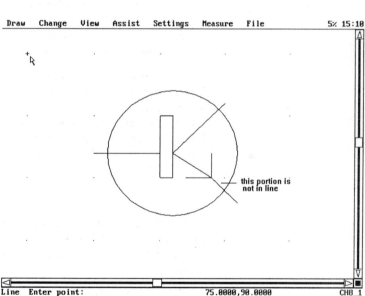

Figure 8.14 *A zoomed view of a single symbol, obtained from a Zoom box.*

keep these settings undisturbed unless it is essential to change them. The grid dots are the same size in a zoomed view as they are in a normal full page view. Always restore the settings after making any alterations to a zoomed view.

☐ You do not know the precise scale of this zoom, but this is not important because you will normally zoom back again.

Zoom Limits is used to restore the screen to the drawing limits that have been set from the *Settings* menu. Use it when new limits have been set, so that the full screen area represents the new limits, and also to restore the full view when you have been careless about zoomed views and cannot recover your original view.

Zoom Full is required when a drawing has used space outside the limits that were originally set. When *Zoom Full* is used, the drawing will be reduced in scale so as to fit into the screen space. This solves the problem about a circuit diagram which, using the standard components, cannot be fitted into the screen space. Using *Zoom Full* on a drawing that is smaller than the limits will make the drawing fill the page.

You can use the *Pan* facility, either by way of the scroll bars, or from the *View* menu, to see detailed views of a drawing which extends beyond the screen limits. This might be because the drawing has been made larger than the limits that were set, or because a *Zoom Box* has been used, and another part of the drawing needs to be inspected in detail without zooming back.

To save time, you can use a zoom box covering more than one symbol shape. What you are looking for in a zoomed out view are such items as unwanted lines, bad intersections (lines crossing instead of a T-junction) and so on. This should all be done *before* any groups are created, because once a shape has been grouped you cannot use actions such as *Line Break*.

For example, the transistor symbol in Figure 8.14 shows that the emitter line is bent (because the emitter was drawn as separate lines with *Snaps* on). If the shape has been grouped, this cannot be changed until the shape is ungrouped:

1. Click on *Ungroup* from the *Change* menu.

2. Select the object to be ungrouped and click.

You can now sort out the drawing. In an example like this, the use of *Snaps* is the problem, because if the emitter line is drawn between snap points, there is no place where the arrow lines can be drawn – there is no other snap position on the line. The remedy is to draw the whole emitter line as one straight line and then put the arrowhead lines, turning *Snaps* off temporarily, and using *Ortho* on (from the *Assist* menu) to draw these lines.

1. Now draw the arrowhead, Figure 8.15. Using *Ortho* on will ensure that the lines must be either horizontal or vertical. You have to judge the equality of the line lengths, or use *Coords* values to measure them.

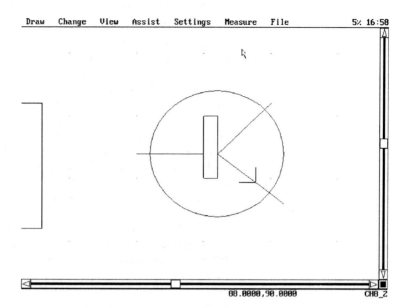

Figure 8.15 *The arrowhead drawn with Snaps off and Ortho on.*

2   Now turn off *Ortho* and restore *Snaps*.

3   Click on *Last View* from the *View* menu to restore the original view of the set of symbols.

This corrected drawing should now be grouped again – it could also be grouped in the zoomed view.

Copy, Move and Rotate

When a symbol shape is zoomed up and corrected, as illustrated, it is not necessary to make the same changes to any other symbol of the same shape which is placed in a different orientation. This applies also when shapes are being created.

For example, the correction to the transistor shape was made on the conventional shape which has the base lead horizontal. It is not necessary to make the same alteration to the transistor symbol which is shown with the base lead vertical, because a copy of the first version can be made and then rotated. The actions of *Copy*, *Move* and *Rotate* are very important in the preparation and use of component symbols.

These commands are found in the *Change* menu. *Move* and *Copy* both make a new copy of an object at another position, but when *Move* has been selected, the original is deleted. To use *Move* or *Copy*:

1   Select the command by pointing to the object (single item or group) or throwing a box round it (a set of lines, not grouped).

2   The message at the bottom of the screen will change to:

    ```
    Copy From point
    ```

    which allows you to move the pointer to some part of the object that you want to locate – usually the end of a lead for a component, but sometimes the centre of the symbol.

3   When you click on this point, the next message is:

    ```
    Copy To point
    ```

    and you should now move the point to where you want the shape to be placed – you will see the shape move with the pointer if you move slowly.

4   Place the pointer exactly where the shape is to be put, and click. *Snaps* should be on if you want to position an object perfectly.

At this last action, if *Move* has been selected, the original shape will disappear, but for a *Copy* there is no change in the original.

Rotation

If you have created a shape, such as the transistor symbol and you want to make another with the base connection vertical, start by copying the original to another part of the screen. You can then Rotate this copy as follows:

1   Click on *Change* and then *Rotate*.

2   You are asked to select the object to rotate, either by pointing or by drawing a frame.

3   You are then asked to click on a centre of rotation – usually this will be the centre of the object, or as near as the use of *Snaps* permits.

4   When you click, you will see a ghosted outline, with a line that can be pulled out by moving the mouse. Rotating this 'handle' line will rotate the image (Figure 8.16).

5   Click again when the rotation is as you want it – make sure that the lines are exactly horizontal or vertical if required.

Rotation allows you to make one master shape and create other versions – though for most circuit drawing purposes you will want only a vertical and a horizontal version of each component shape. You could, of course, create only one version and rotate it as required for a drawing, but this could involve a lot of effort if several rotation actions were needed. By storing a copy of each orientation in a drawing file you can make up drawings, as the following chapter will demonstrate, very rapidly.

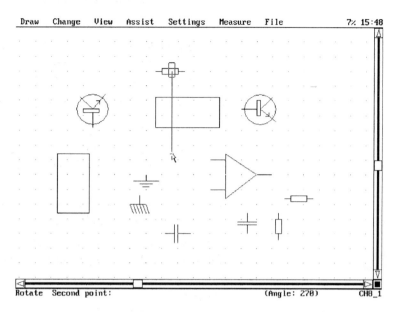

Figure 8.16 *Rotation in progress, showing the 'handle' line that appears.*

☐ Rotation does not allow you to create a PNP transistor symbol from an NPN symbol.

Another useful action is mirroring, to create a mirror image of a component shape. This is particularly useful for transistor and FET shapes. To mirror such a shape:

1 Select the shape to be mirrored.

2 You are then asked, by a message on the bottom of the screen, to select a base point. This is the point about which the object is mirrored, and its selection requires some thought. If you want to mirror horizontally, make the base point to one side; if you want to mirror vertically, make the base point above or below the original (Figure 8.17). A line will appear and can be dragged around with the mouse if you want the mirrored image to be placed at some position other than spaced from the original. Click when the desired position is obtained.

3 The mirror image now appears in its place, and can be moved, copied, etc. like any other object.

218 DRAWING AND CHANGING SHAPES

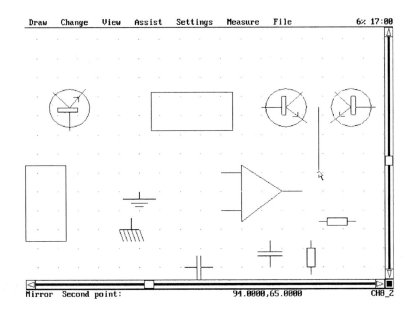

Figure 8.17 *Mirroring with a base point to the right of the original.*

Scale

The *Scale* command, also from the *Change* menu, is used to alter the size of a selected drawing shape. This should not be needed if you have drawn all of your component symbol shapes to the correct scale, but you occasionally find that a symbol of larger or smaller than standard size is needed.

☐ *Scale* needs to be used with care, because altering the size of an object may make parts of it fall between snap points. If this happens, for example, to the ends of the leads for a transistor it will be very difficult to use the symbol in a drawing.

To re-scale an object:

1. Click on *Scale* from the *Change* menu, and select the object whose size is to be changed.

2. You are asked to click on a base point – this should be a point which will be used in drawings, such as the end of a connecting lead to a component.

3. You are then asked to move to a second point. This action will draw a ghosted outline of the shape at a different size, either smaller or larger depending on which direction you move the mouse (Figure 8.18). Try to make this second point also a snap point – never turn off *Snaps* while scaling unless you are forced to.

Figure 8.18 *Re-scaling a component, shown at the second point stage.*

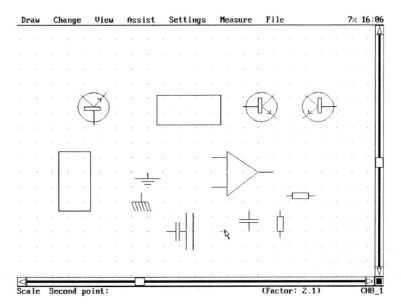

Stretch

The *Stretch* command allows any object to have its dimensions stretched in any direction. Whether the object is grouped or not, Stretch first requires a selection box to be placed, *not* surrounding the whole object. *Stretch* is seldom required in circuit diagram work.

Figure 8.19 *Stretch being used inappropriately on a transistor symbol.*

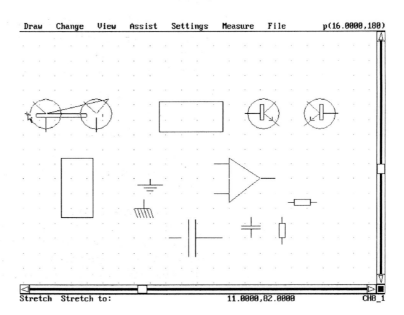

- If the selection box encloses the whole object, a move will be carried out instead of a stretch.

1. Enclose only the parts which are to be moved, leaving the remainder in place, and stretching the lines to fill the space in-between. When the selection box has been drawn, select a point, the stretch base, to move and stretch by moving the mouse to a new position.

2. The image will show the stretch effect, which need not be in a horizontal or vertical plane, it can be in any direction. On components such as transistors, this looks ridiculous (Figure 8.19).

3. When the mouse button is clicked again, the final image will appear stretched to the new size.

- If you want to stretch an object in both dimensions, *Scale* is more suitable.

- IMPORTANT: If any action results in some grid dots disappearing or lines vanishing from a shape, click on *View* and then *Redraw* to make a fresh drawing.

9 Filling and dot creation

Any closed space in a drawing can be filled with a colour or (more usefully for paper printing) a pattern, of which the Solid form is the most useful for electronics circuits. The methods that are used to establish a pattern fill, however, are by no means obvious or straightforward, and can best be illustrated by using two examples, the base symbol in a transistor and a filled dot to indicate a circuit join.

Figure 9.1 shows a zoomed view of a transistor symbol. It is always desirable to carry out a fill action on a view that is zoomed like this because filling involves tracing the outline of the object to be filled, a difficult task on a small view. *Snaps* should be switched on if it was on when the shape was created; in this example *Snaps* is switched on.

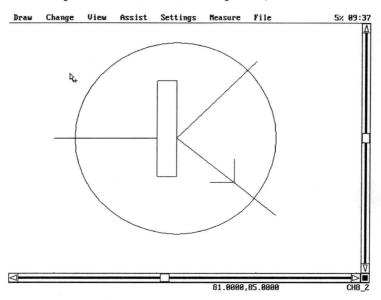

Figure 9.1 *A transistor symbol zoomed up to full screen in order to fill the base rectangle.*

1 Click on *Settings* and *Pattern* to see the *Pattern Settings* menu, Figure 9.2. The default is usually the *Crosshatch* pattern or the *Blank*, so that you will need to select the *Solid* for yourself by clicking on the icon. Do not remove the tick from the box marked *Boundary* unless you

222 FILLING AND DOT CREATION

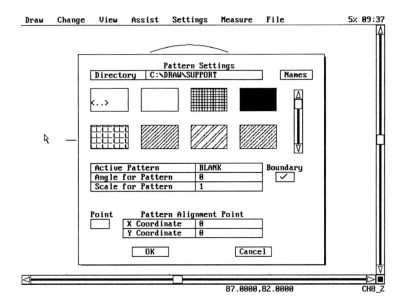

Figure 9.2 *The Pattern Settings menu from which the Solid fill is selected; no other selection is needed from this menu.*

want the boundary lines to disappear once the pattern fill is complete. For circuit symbols, leave this box checked.

- ☐ The *Pattern Settings* menu allows the choice of selection by name or by icon – when icons are displayed you can click on the *Names* box to change the display to names only. When the names display is in use, you can click on the *Icons* box to change to the icons display. Selection is easier using icons because you can see a sample of the pattern.

2 Ignore settings for *Angle*, *Scale* and *Alignment* – they have no relevance to a solid fill.

3 Make certain that the *Fill* item in the *Assist* menu is ticked – click on this line if *Fill* is not ticked.

- ☐ Unless *Fill* is ticked, the filling will not appear.

4 Select *Pattern Fill* from the *Draw* menu, and click at each point on the outline in turn – you will be prompted for the points. In this example, the outline is a box so that you need to click on four points – it is particularly important to click back on the first point again, since this ends the action. After each click, the words *To point* will appear at the bottom of the screen, and the outline of the shape will disappear between the previously-clicked points.

- ☐ This is all much easier if *Snaps* is set on, because it is possible otherwise to click slightly out of position. If you mis-position the cursor when clicking at the first point again, you will have to repeat the action until you hit the correct spot.

5 When the last point is clicked the fill takes place, and you are asked (Figure 9.3) if you accept this filling – the option is to modify it.

☐ Modify in this sense means that you can select another pattern – it does not allow you to repeat the outlining action – you can accept and then use *Undo* if you want to try again.

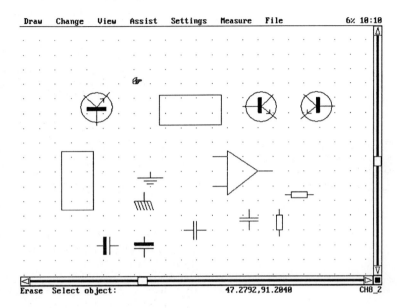

Figure 9.3 *The fill appears, with the option to accept or modify, when the first point in the outline is clicked again.*

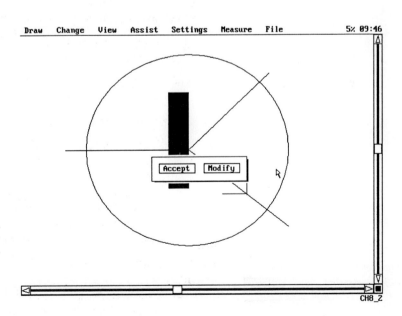

Figure 9.4 *A set of component symbols modified to use filling.*

224 FILLING AND DOT CREATION

Figure 9.4 shows the result of using *Fill* on a set of symbols. In this view, the capacitor symbols have been copied, and a box drawn on one plate of each copy and filled so as to make symbols for electrolytic capacitors.

Filled dots are used either to signify line joins or to show terminals. These require precise and painstaking work to draw by ordinary paper and pen methods, so that the AutoSketch method is an enormous advantage. To create filled dots:

1. Zoom up a blank portion of your drawing that contains component symbols, using a portion that contains four grid dots.

2. Turn off *Snaps* and on four grid points draw circles, each with a different radius.

3. Fill these circles, Figure 9.5. This requires you to click at points around the circles and to preserve the circular shape you will need to use at least eight points, more on a large circle.

☐ You may have problems with the smallest circle. If you find that a fill occurs when you are only half-way round the circle you will have to fill the circle in sections. This action is caused by AutoSketch using an 'almost there' setting for detecting the first point. On a small circle a point just over half-way round is within this radius of detection and is sensed as the first point again. This setting is called the *Pick Interval* and is available from the *Settings* menu. The units of *Pick Interval* are in percentage of screen height, and the default is 1%.

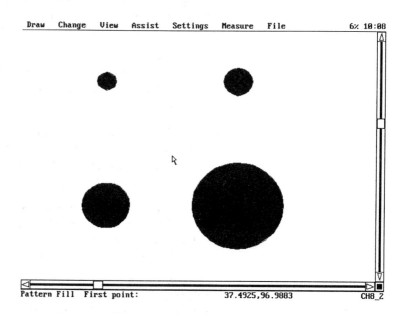

Figure 9.5 *A set of four circles filled so that they can be used as dots.*

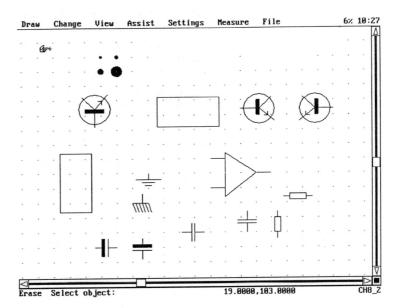

Figure 9.6 *The dots as they appear on the main drawing. Each of these should be grouped. For the next example, they should also be saved using Part Clip.*

4   With the circles drawn and filled, zoom back to the previous view to see the dots on the main drawing, Figure 9.6. Note that the grid dots show up as white dots inside the black filling of these drawn and filled dots.

This provides a set of dots in graded sizes. In the example, the largest pair can be used as line joins and terminal symbols respectively. The smaller dots can be used for other purposes.

- For really precise dot filling, it is an advantage to have the *Arc Mode* clicked on in the *Assist* menu – see Chapter 10. The *Arc Mode* should preferably be used when any circle, ellipse or shape terminated with a curved boundary is to be filled.

- Each dot should be *individually* grouped after filling to ensure that it cannot be broken by careless copying or moving actions. Do not group the whole set as this will make it impossible to select individual dots for copying.

- For filling a curve shape (other than a circle or ellipse), show its straight-line *Frame* to see where to click the mouse. The filling is part of the object and will be erased with the object. To show the frame, click this item on the *Assist* menu so that it is ticked.

As an example of using dots we can look at a set of intersecting lines to which dots can be added. The same scale is used (by loading in the Template file), and in this case there are no component symbols. To get the dots in place, they have been saved as a part (using *Part Clip* from the *File* menu) after creating them in the components drawing.

1. The lines are drawn, and *Part* is selected from the *Draw* menu. This presents a set of drawing files, Figure 9.7.

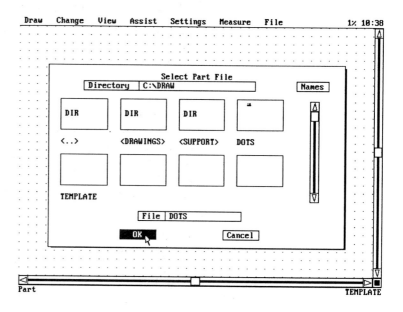

Figure 9.7 *The dots being selected as a part.*

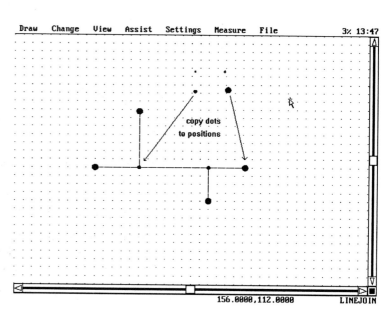

Figure 9.8 *Dots put into place for indicating terminals and joins. The largest two dots in the example have been used.*

2   Click on *Dots* and then on OK to load in the dots. You can then move the cursor and click to place the dots at the cursor position.

☐ If the dot fill has disappeared turn on *Fill* in the *Assist* menu.

3   Now use *Copy* to select dots for the joins and terminals, moving each dot to its place and clicking. The result is shown in Figure 9.8.

☐ Note that dots have been *copied*, not moved. This allows you to keep the dots in hand in case you need to add more dots later as the drawing is developed.

Other patterns

The other patterns are intended mainly for architectural and structural drawings, but some of them can occasionally be of use for electronics purposes. They are labelled by name, including examples from the U.S. standard ANSI and AR sets. Each can be seen in miniature view when you use the (default) *Icons* option in the *Pattern* menu.

CROSSTHCH	SOLID	ANGLE	ANSI31	ANSI32
ANSI33	ANSI34	ANSI35	ANSI36	ANSI37
ANSI38	AR-B816	AR-B816C	AR-B88	AR-BRELM
AR-BRSTD	AR-CONC	AR-HBONE	AR-PARQ1	AR-RROOF
AR-RSHKE	AR-SAND	BOX	BRASS	BRICK
CLAY	CORK	CROSS	DASH	DOLMIT
DOTS	EARTH	ESCHER	FLEX	GRASS
GRATE	HEX	HONEY	HOUND	INSUL
LINE	MUDST	NET	NET3	PLAST
PLASTI	SACNCR	SQUARE	STARS	STEEL
SWAMP	TRANS	TRIANG	ZIGZAG	

Arrays and the inductor symbol

An array is a set of identical objects, and AutoSketch allows two types to be created: box arrays, in which the objects are arranged in rows and columns, and ring arrays, in which the objects are arranged around a circle. Arrays provide a very useful way of making a large number of repetitions of a shape, with pre-determined distance between patterns. The example in this case is of the more useful type (for circuit work), the box array, being used to create an inductor symbol.

1   Start by zooming up a blank part of the components drawing as before. Choose an area to zoom which will be about the depth or length that you want for an inductor symbol.

2   Draw an arc whose ends are two snap points apart, Figure 9.9.

228 FILLING AND DOT CREATION

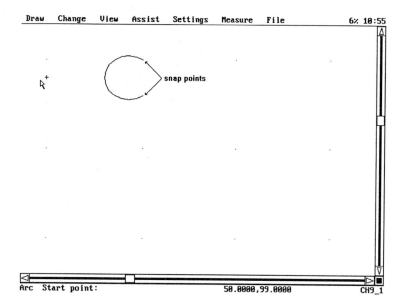

Figure 9.9 *An arc drawn so that a box array can be used to generate an inductor symbol.*

3   From the *Settings* menu, click on *Box Array Settings*, and alter these, as shown in Figure 9.10, to 1 column with a column distance setting of zero.

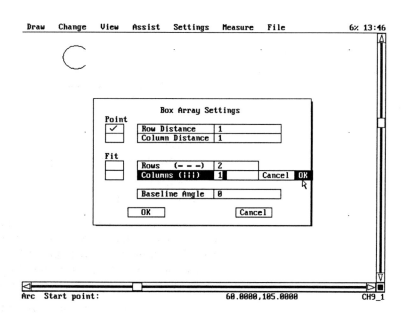

Figure 9.10 *Box array settings.*

☐ To make a vertical inductor symbol requires only one column, and using six rows allows the arc shape to be repeated six times.

4. Now click on *Box Array* from the *Change* menu. You will be reminded that this array has only one column, and asked to click on the object. Click on any part of the arc.

5. You are asked to click on the *First row* point. Click on the top end of the arc.

6. The message is now *To point*. Move the cursor to the other end of the arc. You will see a copy of the arc shape follow the cursor, allowing you to place the cursor precisely, Figure 9.11.

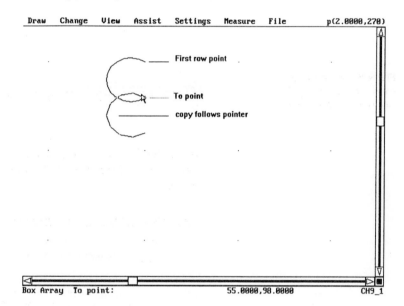

Figure 9.11 *Placing the first copy for the array. This is an important action because it determines how the rest of the array will look.*

7. Click to create the array – you can then opt to *Accept* or *Modify*. This produces the array, Figure 9.12.

8. Zoom back to the previous view to see the inductor symbol appear in its true relative size among the other components.

When you select a box array, the default is to make a small array (one more row and one more column). For larger arrays, select *Box Array* from the *Settings* menu and specify the array by typing row and column separation in numbers, or by pointing with the mouse (tick the *Point* box to enable this, useful for overlapping items).

☐ You can also opt to fit a specified number of rows and columns with the objects, an easy way of creating large arrays. Altering the baseline angle will tilt the objects with respect to the horizontal.

230 FILLING AND DOT CREATION

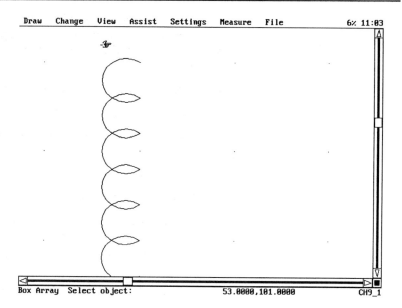

Figure 9.12 *The inductor shape created as an array.*

Remember that dimensions for arrays are in terms of the drawing unit. For example, if the spacing of grid line is 0.2 drawing units, then an array dimension of one grid spacing is typed as 0.2.

Ring arrays are used to draw a set of objects arranged in a circle. This requires you to specify *either* the number of objects in the ring *or* the angle between them (not both). You can type the co-ordinates of the centre, or select *Point* so that the centre can be selected with the mouse. This action is very seldom required for circuit diagram work.

☐ The objects can be rotated so that one axis will always lie along a radius of the circle, or they can be left as they are so that the objects are always facing the same way with respect to the grid lines. You can select a pivot point, which is the part of the object which will always be placed on the rim of the circle of rotation.

Varieties of lines

Thick lines and dotted lines are sometimes required in circuit diagrams, and though they can be created in AutoSketch, they require some planning and some compromises.

AutoSketch 3 has no provision for drawing lines of different thicknesses other than by using polylines. This is an unfortunate restriction, because it can make some actions quite difficult. The ability to use lines of a range of different thicknesses is a considerable advantage, particularly for dotted lines drawn with a laser printer (see later).

The way of drawing a thick line in AutoSketch is to specify a wide polyline of one drawing unit thick. A polyline is normally specified as a thin polyline, and to make this change you need to select the *Settings* menu.

1. Click on *Polyline* in the *Settings* menu.
2. Alter the polyline width figure from its default of 0 to 1.

☐ You can use only whole numbers – a setting of 0.5 will *not* have any effect.

3. Make sure that *Solid Fill* is ticked.
4. Click on the OK boxes.
5. Now use the *Draw* menu and click on *Polyline*. Draw a line in the usual way, but click twice on the last point. This produces a thick line, Figure 9.13.

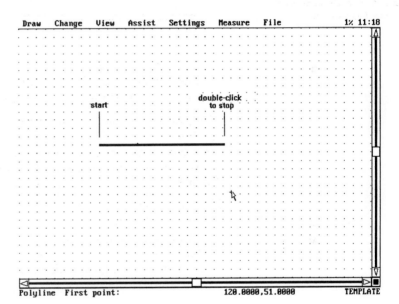

Figure 9.13 *The thick polyline shape. The width must be at least one drawing unit, and only a whole number of drawing units can be specified.*

The snag is that this line may be thicker than you want. Using this type of polyline without snaps makes the line look slightly thinner, but the best way of obtaining thinner lines is to use more drawing points in a drawing, by using 400 x 300 in place of 200 x 150, for example. This is not exactly welcome if you have already created all of your component shapes to suit a 200 x 150 grid. It would be easier if you could opt for a set of multiples of the ordinary printer line thickness.

The default type of line, for use with the Line or other shapes, is a solid line, but a variety of line types is available from the *Settings* menu.

232 FILLING AND DOT CREATION

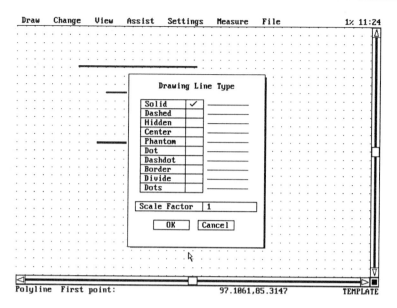

Figure 9.14 *The Line Settings menu – the different effects are not visible with the default scale factor.*

1. Click on *Settings* and then on *Line Type*. This produces the menu, Figure 9.14.

2. The lines of this menu will usually look identical. This is because the Scale Factor setting makes the dots or dashes so small that each line appears continuous.

3. Try a larger *Scale Factor* setting – Figure 9.15 shows a setting of 10 used. At this setting the different line types can be clearly seen.

☐ Do not attempt to use different line styles unless the differences can be seen in this menu.

4. You can now select a line type and click on OK. All lines that you draw will be in this style until you change the *Line Settings* again.

☐ Remember that the effect of zooming will be to alter the spacing of dots or dashes in these different line styles.

One important point, however, is that you may not see dotted lines clearly in a printout, particularly if you use a laser printer. When you set up AutoSketch for use with a laser printer you will normally be asked to select the resolution of the printer in terms of dots per inch – the usual selection is 75, 150 or 300 dots per inch. Most users will want to use 300 dots per inch, the limit of conventional laser-printer resolution.

The problem is that this will result in all lines being printed as $\frac{1}{300}$" thick. This is reasonable (though not easy to photocopy) if all the lines are solid, but a dotted line with each dot only of $\frac{1}{300}$" diameter is almost invisible.

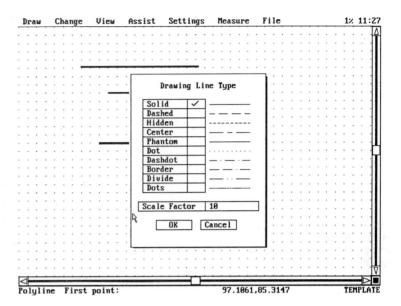

Figure 9.15 *Using a different Scale Factor to make the dotted lines visible.*

- If you intend to use dotted lines, or if your drawings are to be photocopied, it is usually better to set the printer resolution at 150 dots per inch. The snag is that shapes such as circles look more jagged at this lower resolution. Once again, this is a problem that could be overcome if AutoSketch permitted lines of different thicknesses to be drawn but still retaining the highest resolution of the printer.

- Curiously enough, the old low-cost Generic CADD Level 1 CAD program allowed different line thicknesses, but this facility was not put into AutoSketch when AutoDesk bought Generic CADD.

- See Chapter 7 for details of altering the printer resolution settings, either by deleting the SKETCH.CFG file or by starting with the SKETCH3 -R command.

Making a simple drawing

It's time to look at a method of actually constructing a drawing – assuming that a set of component symbols have been drawn, all to the same scale – using the methods that have been outlined in the course of these chapters. The components should preferably be packed in a corner of the screen, Figure 9.16.

AutoSketch is flexible. If you have created a set of components on a screen of 200 x 150 and you want to change your limits so as to use

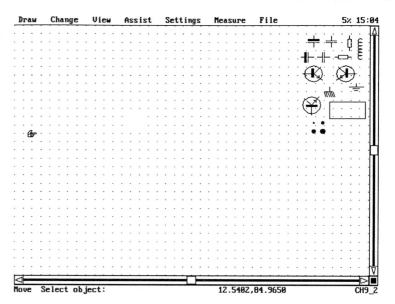

Figure 9.16 *Component symbols set in a corner of a drawing to make room for the circuit diagram.*

400 x 300 (allowing the use of a thick-line polyline), you can change the limits and then use *Zoom to Limits*, altering *Grid* and *Snaps* to suit. The original set of components can now be moved to one corner, but you can either:

(a) Work with them as they are, saving this set as your template

or:

(b) Re-scale the components

Of these options, it is much better to work with the existing components, using zoomed views. The re-scale action does not permit you to scale to a precise amount, such as x2, and will almost inevitably result in the snap points being in the wrong places.

Taking as an example a simple two-transistor circuit, this is constructed as follows:

1 Make sure that *Snaps* is on. Starting with the components template, copy the transistor symbol twice into the vacant space, Figure 9.17.

☐ Remember that once you have select *Copy*, you do not need to reselect if you want to copy more than one object – the *Copy* command can be repeated until you need to use another command.

2 Now copy resistor and capacitor symbols into approximately the correct places. The electrolytic capacitor symbol at the left-hand side has been copied, mirrored, and the mirror image moved into place, with the copy erased, Figure 9.18. The first two resistors will need to be adjusted to avoid their connecting lines meeting (as shown, they make a cross join rather than the standard staggered join).

MAKING A SIMPLE DRAWING 235

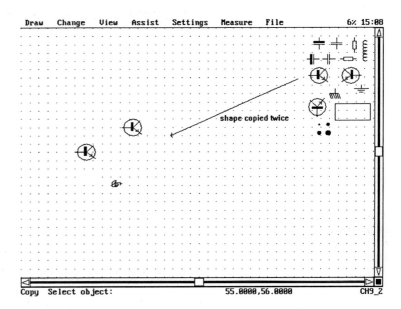

Figure 9.17 *The transistor symbol copied twice – remember that the copy action needs to be selected only once.*

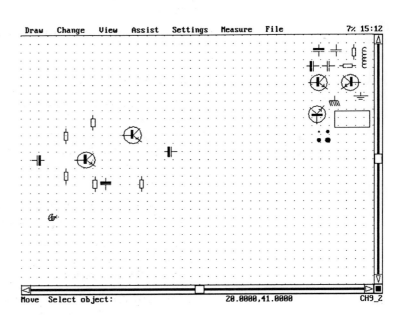

Figure 9.18 *The passive component symbols added in their approximate positions. The resistors on the left need adjustment. Note the mirrored electrolytic capacitor.*

3 Draw the lines that represent the power supply and earth lines, and also the input and output lines, Figure 9.19.

☐ These should be drawn between grid points. Use *Ortho* on (*Assist* menu) if you have any problems in drawing precisely horizontal lines.

4 Now join on the other lines, Figure 9.20.

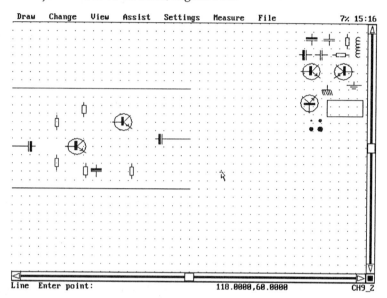

Figure 9.19 *The main lines added to the drawing. Use Ortho to ensure that these are horizontal.*

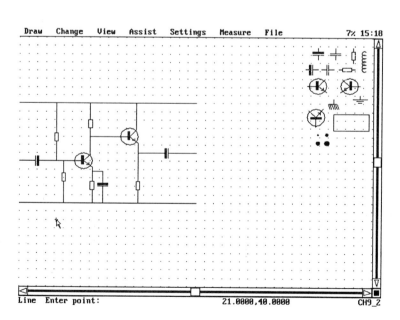

Figure 9.20 *The connecting lines added, also using Ortho.*

5   Add the dots to show joins and terminals, Figure 9.21.

☐   It is important that the centre of each dot should be a snap point, even if the dots were created with *Snaps* off. If the dots are not centred on a snap point, zoom on a dot, switch *Snaps* off and select *Move*. Select the dot and place the pointer at its centre when asked for the point to move. Switch on *Snaps* and select a snap point when asked for a point to move to. This will centre the dot on a snap point.

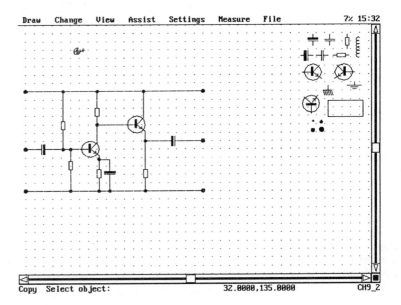

Figure 9.21 *The dots put into place.*

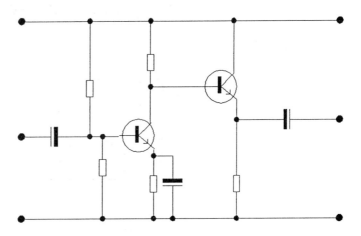

Figure 9.22 *The circuit as it appears when printed using a resolution of 150 dots per inch (for thicker lines).*

The drawing can now be checked and tidied up if needed, using *Zoom Box* to inspect any doubtful areas. Figure 9.22 shows how this drawing will appear when printed using a laser printer. This printout is always noticeably better than the screen appearance (unless the laser printer has been set to a very low resolution).

We now need to look at how to save a drawing as a disk file, and the methods of printing drawings. Once a drawing is saved it can be loaded and edited to any extent you require.

- ☐ If you want to keep an original drawing after editing, you can save the new copy under another filename, using the *Save As* option. In addition, when a file is altered, the original is held on the disk, using the same main filename but with the extension BAK. If you want to re-use a file which has become a BAK file, use the MS-DOS RENAME action to change the name to something like TEMP.SKD so that you can use it. You cannot rename a file to a name that is already in use, so that if you have LIN2IC1.SKD and LIN2IC1.BAK, you will have to rename LIN2IC1.BAK to something like TEMP.SKD in order to use it.

Saving files

When a file has been created using the *New* option from the *Files* menu and starting from scratch, or when AutoSketch has been started afresh, the file will be un-named and all options will be at their default values. The *Save* option of the *File* menu is intended to deal with an already-named file, so that *Save As* is a better option. *Save As* also allows you to change the name of a file. This allows you to load a file, alter the drawing, and then save under another name, a very useful process. Get into the habit of saving a file with *Save As*, because this reduces the risk of erasing a valuable file when you have made some alterations to it and saved it using the same name.

- ☐ The filename for a drawing is always displayed at the bottom right-hand corner of the screen. For a new file, this spaces shows 'Unnamed'.

Using Save As

When you take the *Save As* option, you are presented with the Save box over the drawing, Figure 9.23.

1. Place the cursor into the name box, and type a suitable filename, obeying the usual MS-DOS rules on filenames (8 characters maximum, starting with a letter).

2. Confirm that this is suitable by clicking on the OK box.

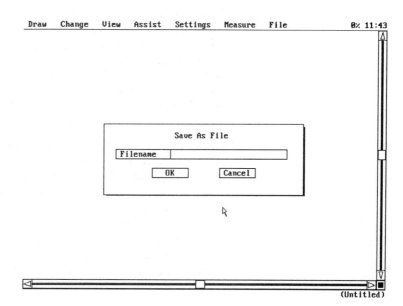

Figure 9.23 *The Save As box – you can type in another filename to allow the file to be saved under another name.*

3   Save the file by clicking on the other OK box at the bottom of the *Save As* box. You are given two chances to Cancel if you change your mind, and following the *Save As* action the drawing remains on screen undisturbed.

Using Save

Save should preferably be used when a filename already exists and you do not need to change it. If you opt for *Save* using a new file, however, you will see the *Save As* file box appear, requesting a filename. In either case, type a name of up to eight characters, and the file will be saved along with the other AutoSketch files. You can also include the drive and/or directory in the name to save the file elsewhere.

☐   All drawing files are saved with the SKD extension letters – you do *not* type these. For example, a file that you save with the typed name of LIN2IC will be saved under the name LIN2IC.SKD.

Loading a file

When a drawing file has been saved, you can recover it on screen by using the *File Open* action as follows:

1   Select *Open* from the *File* menu. You will see a box containing icons (miniature drawings) and names of files, in alphabetical order, that have previously been created.

240 FILLING AND DOT CREATION

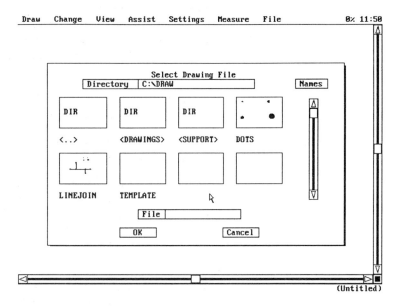

Figure 9.24 *The Open File display, showing directory names as well as filenames.*

☐ Other directories are indicated by the name placed between angle brackets, Figure 9.24. Click on a directory icon to obtain that directory.

2 Use the scroll bar in the *Select Drawing* box to display another set of icons and filenames if the file you want is not in the first set.

☐ You may have placed the file in another directory.

3 Select the file you want by placing the cursor on the icon and clicking. This places the filename into the box marked *File*.

4 Click on OK to confirm and so load the file. You can click on Cancel if you decide to use another file.

☐ Note that saving a portion of a drawing as a part requires you to use *Part Clip* from the *File* menu for saving, and *Part* from the *Draw* menu for loading.

Printing or plotting

AutoSketch distinguishes between *printing* and *plotting* for the methods of printing out the drawing. For the best possible results on line drawings, a pen-plotter should be used, but a good second-best is a laser or inkjet printer, and a dot-matrix printer can prove adequate for some purposes. We shall concentrate on the *Print* options, because most users of AutoSketch are likely to use a printer.

The Printing commands

Selecting *Print* will print out the entire drawing on the selected paper size; you need only ensure that the printer is ready. This is by far the simplest method of printing, but it is not necessarily the most satisfactory.

- The drawing will totally fill the width (or depth) of the paper, and some lines at one or both edges may be broken.

- You may not want the drawing to fill the width of the paper but be confined to a part of the paper.

To print only part of a drawing, or print on only a part of the paper, you must create a *plot box*, which surrounds the drawing. Only the portion of the drawing that is included inside the plot box can be printed, and though several plot boxes can be created, only one can be used for printing. The plot box is created from the *Print Area* menu.

- The plot box is itself a drawn object so that it can be selected, scaled and moved like any other object on a drawing. This provides a very flexible way of dealing with printing, which for most purposes is superior to using the plain *Print* command. You can even keep different plot boxes on different layers (see Chapter 10) so that the same drawing can be printed at different scales.

- A drawing can be saved with its plot box to allow the same drawing to be printed again in the same way, or to allow the plot box to be further edited later.

The creation of a plot box is started by clicking on *Print Area* from the *File* menu. This brings up the *Plot Box* menu, Figure 9.25.

1. Select *Paper Size* for one of two fixed paper sizes, the USA size of 8.5" x 11" or the A4 size of 297mm x 210mm; or fill in your own values. The USA size uses plotting units of inches, and A4 uses plotting units of millimetres. Unless your *Drawing Limits* have been set to the same as the paper size, you cannot use a 1:1 ratio of *Drawing Units* to *Plotting Units*. The US size is the default, and when you click on A4 the units will change to Millimetres (the US spelling – *millimeters* – is used)

☐ If you are using a paper size other than the two shown, fill in values for X and Y sizes (allowing a margin) in the spaces provided. You will be reminded not to fill in sizes greater than your plotter/printer can cope with.

☐ If the X size of the plot is greater than the Y size, it is better to rotate the plot so that it is printed sideways on the paper. Click on the *Rotate by 90 degrees* box to change Off to On.

242 FILLING AND DOT CREATION

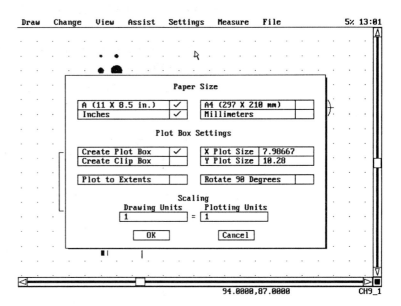

Figure 9.25 *The Plot Box menu, showing the defaults and allowing you to fill in settings.*

2   Select scaling that will allow the whole drawing (or a suitable part) to be seen. You can move a plot box using the *Move* command – see later – to cover a desired part of a larger drawing. In all cases, what is seen inside the plot box is what is plotted.

☐ You can also opt in the *Print Area* selection to create a *clip box*. This is another way of selecting a part of a drawing – only the portion that is inside *both* the plot box and the clip box will be plotted. This can be useful if, for example, you have left a set of component drawings on a page occupied mainly by a circuit diagram.

☐ The *Move*, *Scale* and *Stretch* commands of the *Change* menu can be used on both types of boxes.

3   Click on OK to end selection. If you cannot see the plot box after completing the *Plot Area* page, select *Last Plot Box* from the *View* menu.

This outlines the general ways in which the plot box commands are used, but a lot of time can be wasted in trial and error until you are accustomed to the effects of the commands. A general method which is rapid and efficient is as follows:

1   Click on *Print Area* as usual, and click on *A4* in the *Paper Size* section. This will automatically select millimetres as units.

2   Click on *Print to Extents*. This will ensure that the plot box covers the whole drawing. Click on OK.

PRINTING OR PLOTTING 243

3 When the plot box is ready, click on *Accept*. Click on *View* and on *Zoom Limits*.

4 You will now see the plot box in part of the screen, Figure 9.26. Click on *Change* and on *Scale*.

5 Re-scale by pointing to the plot box, clicking and then dragging out a larger plot box, Figure 9.27. Click when the size is appropriate.

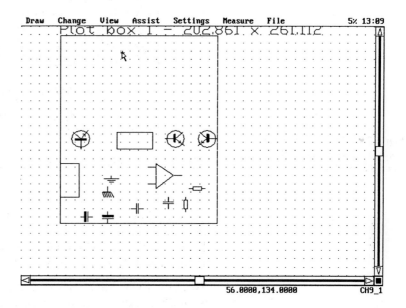

Figure 9.26 *The plot box after zooming to limits. Note that some shapes are hard against the edge and will not print satisfactorily.*

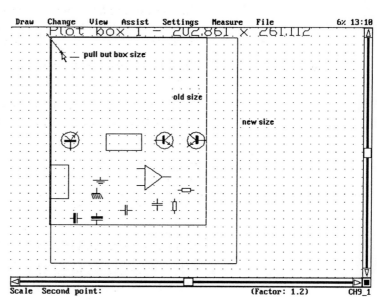

Figure 9.27 *Dragging out a larger plot box. This can then be moved to centre the drawing in the box.*

6 Click on *Change* and then *Move*, point to the plot box and then move it so that the drawing is centred.

☐ If the plot box is too large so that it fills the screen, use a *Zoom X* action, filling in a *Magnification* of 0.5.

7 When the appearance of the drawing in the plot box corresponds to how you would like to see the drawing in the page (centred, at the top or bottom, sideways, etc.) use the *File* menu and select *Save*. When this has been done, switch the printer on and click on *Print*.

This method will always result in a satisfactory print, something that is not assured by using the plain *Print* command or by filling in values on to the *Plot Box* form.

10 Text, layers and other items

Adding text

Text, as added to circuit diagrams, consists normally of small portions of letters and numbers used in circuit references, such as values of components or their circuit references: R27, C52 etc. AutoSketch makes two forms of provision for text, one called *QuickText* for short single line text; the other, *Text Editor*, for longer multi-line pieces of text. For the purposes of circuit diagrams, only the QuickText is of real interest. This is selected from the *Draw* menu by clicking on *QuickText*.

The features of text that are important for the purposes of lettering circuit diagrams are the text size and its style or *font*. There is a huge range of sizes available because text consists of drawn shapes and can be scaled, copied, moved, mirrored etc. like any other AutoSketch object. The number of fonts is more limited, and the default is a font that looks rather like hand-lettering. More elaborate fonts can be used, though when a drawing contains a large amount of text in an elaborate font it can take a considerable time to re-draw completely.

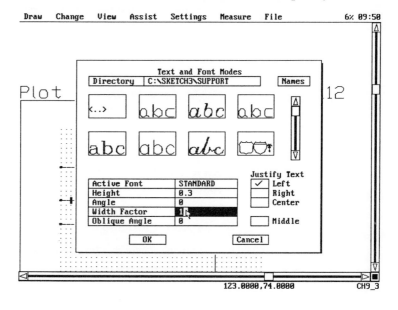

Figure 10.1 *The fonts and character size/style options of the Text settings.*

☐ Remember that if you do not need a total re-draw you can stop the action by pressing the Ctrl-C keys. This only stops screen re-display, it has no effect on the drawing stored in memory.

The default size of text characters is related to the drawing unit, and unless you are working with the defaults, the text size will always need to be changed. Several font sets of characters exist. The default font is the simplest, providing characters that are drawn with simple straight-line strokes. If you want any other fonts, you have to load them in, using the *Settings* menu. The other fonts include a mathematical set, a music set and a mapping set.

Some of the fonts provided are illustrated in Figure 10.1, obtained by clicking on *Text* in the *Settings* menu. The full set is:

Standard	A simple line font
ItalicC	A more curved and sloping font
Monotxt	A development of Standard
RomanC	A more elaborate font similar to a printed font
RomanS	A good-looking general-purpose font
Symap	A set of symbols for maps
Symath	A set of mathematical symbols
Symusic	A set of symbols for writing music

☐ Other fonts can be added – AutoSketch for Windows possesses a larger set of fonts, and fonts can also be bought separately, as can packs of symbols for specialised purposes. Text can be selected, cut and pasted, and the Text Editor can be used like a word processor. Remember that text will look better on paper than it does on screen.

Text settings

The text options of the *Settings* menu, other than font selection, are *Height, Angle, Width Factor, Oblique Angle* and the *Justification* settings.

For most applications using the *Standard* font, the only alterations needed in the *Text* and *Font Modes* menu will be to alter sizes and angles. The best initial option for *Height* is one grid spacing, whatever this happens to be set to. You may want to vary this when you see the text, but it makes a good starting point. The default of 0.3 units is often too small.

☐ You can reduce the *Width* setting to cram more text into a given space, or increase the setting to spread the text out.

☐ *Oblique Angle* can be used to emphasise words or phrases, making the text appear in italic style.

The *Angle* option is used to make the text line up with lines that are not horizontal. In practice, it is better to have text either vertical or horizontal, not at a mixture of angles, and for circuit diagrams it should never be necessary to use text that is printed on a slanting line.

Using other fonts

To select a new font from the icons, look for the icon that you want – this shows a sample of the font. Use the scroll bar to view a different part of the font set if the font that you want to use is not visible. Click on the icon and then on OK to load the font. You can experiment with width and oblique angle for the different fonts – reducing the width makes some fonts appear much more readable on the screen, and a small change of oblique angle can sometimes be beneficial also.

□ Always make a note of settings if these are not defaults. This can be done by placing text outside the print box area, or otherwise well away from the drawing. If you save a file that contains unorthodox font settings these settings will be saved with the file and will be available when the file is loaded so that you can edit using the same settings.

Typing the text

For circuit diagrams, always use the *QuickText* option from the *Draw* menu to make the text appear on the screen, after setting the height of the text in the *Settings Text* menu. Place the cursor with the mouse at the place where you want the text to start, and click to establish the starting position. Type a line and then press ENTER to end typing. If you need a new line before you end the text entry, click the mouse on the new line *without* pressing ENTER.

□ The alternative for large amounts of text is to use the Text Editor. You will not then see your text appear in a font style until it is placed in the drawing. The Text Editor is not necessary for most circuit drawing applications, other than lists of components: see below.

There are several special effects that use the %% signs to start and stop effects. These are:

%%o	Overscore
%%u	Underscore
%%d	Degree sign
%%p	± symbol
%%c	Diameter dimensioning symbol
%%%	Single percent

248 TEXT, LAYERS AND OTHER ITEMS

As an example, place some text on the sample circuit diagram, using the RomanS font. This is done as follows:

1. Use *Text* from the *Select* menu to choose the *RomanS* font and alter the *Height* setting to 2 units.

2. Zoom the drawing up to full screen size, using *Zoom Box*.

3. Click on the *Draw* menu and then on *Text*. Place the cursor above the first capacitor symbol and click. You will see the text pointer or cursor appear – a horizontal line whose size shows the size of the text. Type the reference C1. Press the ENTER key to fix this text.

4. Continue by placing the cursor near the next component, clicking, typing text, and then pressing the ENTER key.

The result is shown in Figure 10.2, showing the zoomed screen view. As always, the printed version will look better than the screen view.

☐ Each line of text is treated as a separate object and can be moved, copied, scaled and so on like any other drawn object. You cannot, however, separate items of text on the same line. For example, the C1 entry cannot be separated into C and 1.

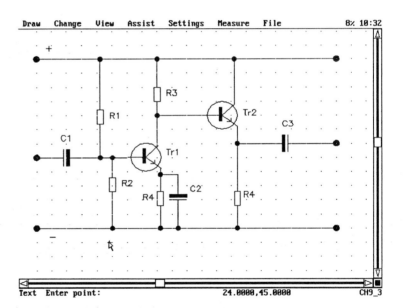

Figure 10.2 *A zoomed view of the circuit example with components labelled using the QuickText method.*

Using the Editor

The Editor is rarely needed for straightforward labelling of circuit diagrams, but it can be useful for creating tables or lists of components. In a crowded diagram it can be useful to label each component with a reference such as R22, C56, and to show the component values separately.

☐ This is not necessary if drawings allow space for component values as well as reference numbers.

The editor is started by clicking *Text*:

1. You will be asked, by a message on the bottom line of the screen, to enter the point where text is to begin. Place the point on the chosen place and click.

2. The Text Editor screen appears, Figure 10.3. Click on *Settings* first to ensure that you set the size and font for your text, because you will not see the actual text appear until you have finished with the Text Editor.

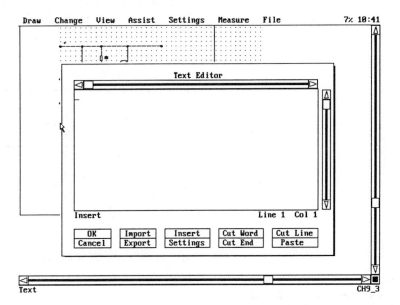

Figure 10.3 *The Text Editor screen. This should be used for tables and other larger pieces of text.*

☐ The settings that may already have been made for use with Quick-Text do not affect the Text Editor, which sets text properties only from within the Editor.

3. Text is typed in the main box, which is provided with scroll bars to allow text to take up a space that is larger than the provided area. The pointer or cursor for text is a short horizontal line. Text does

not appear in the font that has been selected, but in a standard form of 'machine font'.

4   As you type the text you can alter it using the normal methods of word processing; see below.

5   Text is placed into the drawing when you click on OK. Clicking on Cancel results in nothing being placed on the drawing.

If the text that appears is in the wrong size or font you can change the text properties as follows:

1   Select *Settings* and click on *Property*. Make sure that *Font* and *Text* are ticked.

2   Also from the *Settings* menu, click *Text* and set the font and size you want to use.

3   From the *Change* menu, click on *Property*. Select the text you want to change by dragging a selection box around it.

4   When you click on the selection box second point, the text will be changed – this can be a slow process.

The Editor offers some refinements that cannot be used with the QuickText method.

1   Clicking on *Insert* allows you to type using insertion (the default) or overtyping (click to change from one to the other). When *Insert* is used, a character typed at an position in an existing line will cause characters to the right of it to move right so that the new character is inserted. With *Overtype* selected, a new character will replace the one that the cursor is over.

2   *Cut Word* will delete the whole word or part of a word starting from the cursor position.

3   *Cut End* will delete everything from the cursor position to the end of the line.

4   *Cut Line* will delete all of the line that contains the cursor. If there are lines above and below the cut line, these will be moved to close the space.

5   Whatever was last cut (word, phrase or line) can be restored at the cursor position by clicking on *Paste*. You can use this to move or copy text from one position in a piece of text to another.

☐ Once text has been placed on a drawing, you can only return to edit it by using the *Text Editor* option from the *Change* menu. The process is described below under 'Exporting text'.

USING THE EDITOR 251

The Editor also allows you to import or export text. Text can be imported from another source, usually a word processor, only if the text is in the standard form called ASCII. Any program described as a text editor will produce such text – this includes the Amstrad RPED program, the Edit of MS-DOS 5, the Notepad of Microsoft Windows and many others.

☐ The complication here is that, though any word processor will save text in ASCII form, different word processors use different names for these files. For example, WordStar describes ASCII files as 'Non-document files' and WordPerfect refers to them as 'DOS files'. There are several utility programs (see the PDSL list) that will convert word-processed files from their native form into ASCII, and some word processors incorporate such utilities.

Import is used mainly if you have some *short* standard text, such as company name and address, disclaimers, reminders and so on that will appear in every drawing. To use *Import*:

1 Click on *Import*. You will see the *Import* box appear, Figure 10.4.

2 You need to know the name and file path for the file you want to import. There is no option to search through directories as you have in the main drawing files.

☐ The maximum length of file that can be imported is 2048 characters.

3 The imported text can then be edited as required and placed on the drawing by clicking on OK.

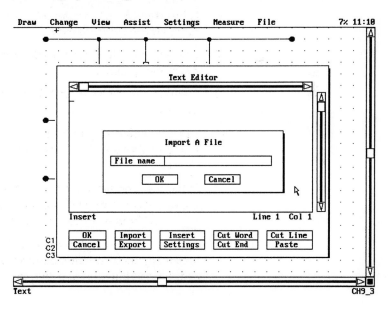

Figure 10.4 *The Import box of the Text Editor. This allows you to import short ASCII files from other editors.*

Text can also be exported, but only if it is visible in the Text Editor (this excludes QuickText). If you have just typed some text it can be exported immediately, but if you want to export or edit text that has been placed on the drawing:

1. Click on *Text Editor* in the *Change* menu.

2. When the message at the foot of the screen reads 'Select Object', click on the existing text. This will then appear in the Text Editor.

3. You can now change or export this text. If you want to export the text you must type a filename and (if necessary) path.

Summary of editing keys	
Key	**Action**
Backspace	Moves left and erases character to the left of the cursor
Del	Erases character at cursor, text moves left to fill gap
Right cursor	Moves cursor one character right
Ctrl-Right	Moves cursor one word right
Left cursor	Moves cursor one character left
Ctrl-Left	Moves cursor one word left
Up cursor	Moves cursor up by one line
Down cursor	Moves cursor down by one line
PageUp	Moves cursor up by one text page
PageDown	Moves cursor down by one page of text
Home	Moves cursor to first character in text

Mu and ohm signs

The mathematical signs font of AutoSketch does not include the *mu* and *ohm* signs that are used in circuit diagrams. In modern practice, the ohm sign is less common because diagrams conforming to the B.S. drawing standards will use terms such as 1R5, 2k2, 3M3 and so on to indicate resistor values without using the ohm sign. The mu sign is, however, needed for capacitor values that are not in the nF or pF range.

These signs can be created and added to your page of standard symbols, but some care is needed to ensure that the scale is correct. The following method is suitable:

1. Using QuickText, place a letter 'u' and letter 'o' on a small portion of the page, using a suitable scale size and the font that you use for your other characters. Press ENTER after typing each character.

2. Zoom up this portion of the page using *Zoom Box*.

3. Turn off *Snaps*, and add a tail to the u sign to make it look more like a mu, Figure 10.5.

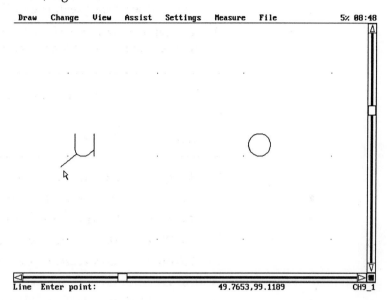

Figure 10.5 *Converting the 'u' into a mu symbol, with Snaps off.*

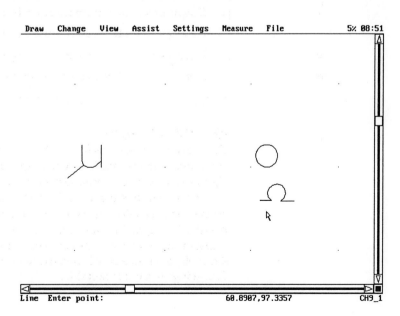

Figure 10.6 *Using Arc and Line to create an omega sign; the 'o' is present only as a guide to size.*

4   Using the 'o' only as a guide, create an arc of about the same size and add two lines, Figure 10.6.

□ Remember that the letter 'o', like any text character drawn by AutoSketch, cannot be broken, so that you cannot create the shape by breaking the 'o'.

5   Now group each symbol separately, move the symbols closer to each other, placing the moved symbol on a grid point, and turn on *Snaps* again. Zoom back to the original view.

This will allow you to treat these characters as symbols, copying and moving them into place along with other text. When typing a line such as:

2μ2

use QuickText to type the first 2 and press ENTER. Shift the mu sign up against the numeral, and then, further away, use QuickText to create another 2. Move this 2 back against the mu.

□ This avoids the problems that can be caused if you try to leave a space between the numerals and shift the mu sign between them. The characters appear as squares during a move action; the square accurately indicates the size of a character.

Using layers

So far, we have been creating drawings on a single layer. AutoSketch can work with the screen as if it consisted of up to ten separate transparent sheets, each of which can be drawn on separately. You can then choose which sheets are to be visible and which sheet you are working on.

□ This allows you, for example, to draw component symbols on one layer, interconnections on another, and lettering on a third. The merit of this is that you can alter items in one layer with absolutely no chance of accidentally changing the objects in another layer. You can also produce drawings in which it is easy to print different versions, one lettered only with component reference numbers, another with only component values. Whatever is printed is determined by what is visible at the time.

The use of layers is governed by clicking on *Layers* in the *Settings* menu. This produces the *Layers* menu, Figure 10.7, in which you can see the default selection of all layers visible and Layer 1 current. Your choices are:

USING LAYERS 255

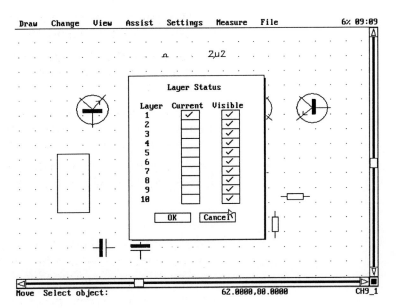

Figure 10.7 *The Layers menu that permits working with different parts of a drawing separately.*

- To work only on a single layer, usually Layer 1, as we have done so far

- To work with one layer but have drawings on other layers visible

- To work with one layer and leave drawings on other layers invisible

Note that you are not allowed to work on more than one layer at a time. You can, however, move an object, a set of objects, or a complete drawing from one layer to another. The method of doing so is not obvious, however, and using *Move* or *Copy* is not the correct method.

1. Use *Layer* on the *Settings* menu to make another layer current. Leave the layers visible (unless there are other layers which would obscure your view).

2. Click on *Property* from the *Change* menu.

3. Select the object or set of objects you want to move to the new current layer. You will need to drag a window around a set of objects, but a single object can usually be pointed to.

4. Check that the move has taken place by making the new layer visible and the old layer invisible. You can make the new layer current if you want to work further on it.

☐ Though this action carries out a move only, there is no reason why you should not copy objects to an unused part of a layer and then

256 TEXT, LAYERS AND OTHER ITEMS

move them to the new layer. In addition, with the new layer current you can use the *Part* command from the *Draw* menu to place objects from the disk files.

See the following paragraphs for a more detailed explanation of properties.

Properties

The *properties* of a drawn object are listed in the *Change Property Modes* menu. This menu is obtained by clicking *Settings* and then *Property* and it is illustrated in Figure 10.8. This list consists of settings that are normally made before an object is drawn, but which can also be altered after the object has been drawn. Of the list, the properties of *Font*, *Layer*, *Line type*, *Pattern*, *Polyline width* and *Text* might need to be altered for objects in a circuit diagram.

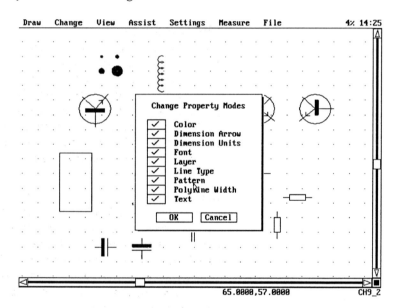

Figure 10.8 *The Property list in the Settings Property menu.*

The *Change Property* list shows all of the items ticked. This means that when an object is selected, any of these properties, as applied to the selected object, can be changed. If there are properties which you would not want to be changed, remove the ticks by clicking on each in turn.

☐ Unless there is any chance that you might alter several properties accidentally, it is just as well to keep all properties ticked. The merit of keeping all properties ticked is that you can alter several properties at once.

The method of altering properties is:

1. Use the *Settings* menu to alter all the settings that you want. You might want to alter line style, for example, or line style and colour together.

2. Also in the *Settings* menu, ensure that the *Property* list has ticks against the items you want to change.

3. Click on *Change* and then on *Property*.

4. Select the object or objects which you want to re-draw with the new settings.

5. The objects will be changed in accordance with the new settings, and you can select others to change until you make some other menu choice.

Minor matters

The topics that have been dealt with in these chapters (7 to 10) are those essential to the drawing of circuit diagrams with AutoSketch. There are a number of features which are of minor importance to the use of AutoSketch for circuit diagrams, but which are important when the program is used for other purposes. A few of these can sometimes be useful, and it is always an advantage to know just how much a program can do; in the following pages these less important (for circuit diagram work) actions will be summarised.

Using frame

Frame means the display of straight lines that are used in creating a curve. When you click on *Curve* from the *Draw* menu, you create the curve by clicking on points that are at first shown joined with a jagged set of lines. These lines are the *frame* of the curve, and they are smoothed out into a curve when you click twice on the last point.

By selecting *Frame* from the *Assist* menu, you can see the frame displayed for any curve. This is useful mainly if the curve forms a boundary to a closed shape that you want to fill, because the fill action requires you to click on the frame points rather than on the points of the smoothed curve.

Arc mode

Arc mode can be switched on (*Assist* menu) while a polyline is being drawn or a closed shape filled. With arc mode on, the boundaries of the polyline or the fill will be arc-shaped rather then straight-line sections.

☐ This can make the filling of circles much easier and of better appearance.

258 TEXT, LAYERS AND OTHER ITEMS

Settings summarised

The following settings are shown mainly for reference – the settings most appropriate for circuit drawing have already been covered in detail.

Arrows
The *Arrows* setting box, Figure 10.9, is used to select the type of arrow that will appear in dimensions for mechanical drawings. The choice is from a set of five styles, including a plain line.

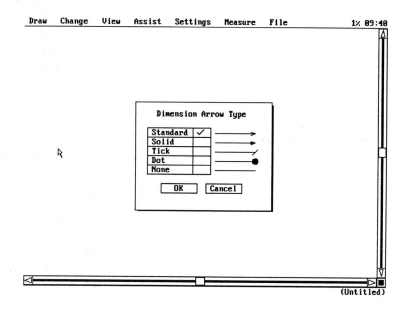

Figure 10.9 *The Arrows setting box for selecting the type of arrows used in dimensions printed on the drawing.*

Attachment Modes
The *Attachment Modes* settings box, Figure 10.10, is used where *Attach* is preferred to the use of *Snaps*, and allows preferred attachment points for objects to be selected. The default is to have all attach points on. The use of *Attach* can also be switched on or off in this menu.

Box Array
These settings have already been demonstrated in Figure 9.12. They allow you to specify the number of rows and columns, and the spacings (either by pointing with the mouse or by supplying distances in terms of drawing units).

Chamfer
These settings are used for machine drawings to create a flat join where two straight lines meet, so that

MINOR MATTERS 259

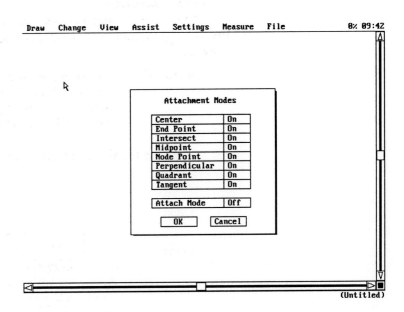

Figure 10.10 *The Attachment Modes menu – useful mainly when you are working with Snaps off.*

there is a flat portion rather than an apex. The default distance apart at which the flat section starts is 0.5 drawing units in each direction.

Color   Allows the drawing colour to be set as Red, Yellow, Green, Cyan, Blue, Magenta or Black. Only Black can be used on a monochrome screen. The colour can also be specified as a number in the range 1 to 127. Only the first seven colour have standardised names, and the ability to use colours depends on the type of display system your computer uses. (Though printing will normally be in black and white, it can be an advantage to use colour for different parts of a drawing to distinguish different sections or purposes, particularly on buses of parallel lines.)

Curve   This setting allows you to specify how many straight-line segments will be drawn when you draw a curve (before the curve is smoothed). You can also specify *Frame* on or off in this menu. You can use a number of segments up to 100. A larger number of segments makes a curve more precise, but drawing takes longer.

Ellipse   Ellipse settings are concerned with the points that have to be specified to draw the ellipse. The default is to click on the centre and one end of each axis. The other options are to click on the centre and the end of one axis and then to move the mouse to locate the other axis; or to click on each focus and on a point on the rim.

Fillet This setting is used to make a curved transition between meeting straight lines. You can select the radius of curvature, with a default of 0.5 drawing units. Compare *Chamfer*, which makes a straight-line transition.

Grid The *Grid* setting, zero by default, is used to specify the grid dot spacing in drawing units. This topic has been dealt with.

Layer Layer setting have been mentioned earlier in this chapter.

Limits This setting is used to determine the normal limits of display. A change of limits has to be followed by using *Zoom Limits* if you want to see the effect on screen. Limits are not restrictive – they simply define the area over which the grid will be placed and there are no restrictions on drawing outside the set limits. Units are drawing units only, and bear no relationship to printed size or to real-life units (millimetres or inches) until the drawing is plotted.

Line type This setting allows the style of line to be determined, as described in Chapter 9; see also Figure 9.14.

Part Base When an object or set of objects is saved as a part, the *Part Base* is the point that will be attached to the cursor when the part is loaded into a drawing. The default method of selecting the *Part Base* is to click on the point, but the settings allow you to specify co-ordinates (seldom useful) or to use a default base point.

Pattern These settings are used for a fill and have been mentioned in Chapter 9. The settings are illustrated in Figure 9.2.

Pick Interval This setting determines how close the pointer needs to be placed to an object in order to select it. The default is 1% of screen height.

Polyline This setting allows you to specify polyline width in order to draw lines of different thickness. The default is 0, and any number that is used must be a whole number of drawing units. See Chapter 9 and Figure 9.13.

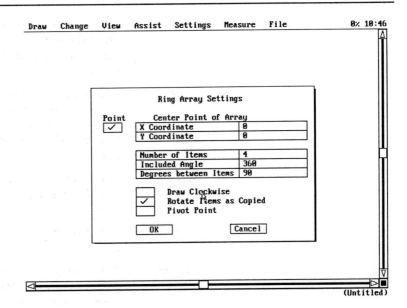

Figure 10.11 *Ring Array settings*.

Property   Allows properties to be selected for changing; see earlier in this chapter.

Ring Array   These settings, Figure 10.11, are used to describe the creation of a ring array, in which copies of an object appear placed in a circular pattern. You can specify co-ordinates of the centre, but the normal default is to point with the mouse and click. The number of items, total angle of rotation and angle between items can all be specified, with the defaults of 4 items, 360° total and 90° between objects. The three check boxes on the menu contain only one default, to rotate the objects as they are copied into the array. With this default on, copying a set of Y characters will result in the tail of the Y always pointing to the centre of the circle. Clicking on this box will result in objects being copied in the same orientation as the original. The default is to draw a ring array anti-clockwise, but checking the appropriate box will cause the array to be drawn clockwise. Checking *Pivot point* allows you to specify a point in each copy which will be placed at a uniform distance from the centre as each copy is made.

Snap   These settings, dealt with earlier, allow snap spacings to be set and snaps switched on or off.

Text   The *Text* settings have been covered in this chapter.

Units   The *Units* settings were noted in Chapter 7.

Measurements and dimensions

Measurements can be of two types – information temporarily put into a box, or data printed as part of the drawing. These all produce the information in a box, which disappears when you click on OK.

1. For measuring *Distance*, click on the points whose distance apart you want to find; the result is in drawing units.

2. For measuring *Angle* click on a base point (one end of a line) and then on two bearings – one of which will be the other end of the line and the other some point whose angle you want to measure. The result will be an angle in degrees, between zero and 180°.

3. For measuring *Area* click at points around the perimeter of a shape; the area is shown when you click at the first point again (which is marked).

☐ Select *Point* to show the co-ordinates of the pointer as you move it (like selecting *Coords*). When the button is clicked, the co-ordinates of this point are displayed in a box. This is necessary only when working in feet and inches, with non-decimal fractions. Select *Bearing* to measure angles to the horizontal when you click on two points of a line.

The *Angle, Align, Horiz* and *Vert* dimension options all produce printed data on the drawing, and you have to indicate (by clicking) where the line of print will appear (it can be moved or erased if needed).

☐ The *Align* dimension measures distances along a line which is at an angle, not vertical or horizontal; otherwise, the meanings are all as for the measurements above. There is a choice of arrows and other dimension indicators.

Once an object has been dimensioned, any action that changes its dimensions (such as *Stretch*) will also change the dimension data provided that this has been included inside the selection box. You need to take care to include the dimension line if a dimensioned object is to be moved or erased.

DXF and SLD files

AutoCAD is the big-brother program of AutoSketch, a fully-featured CAD program whose price reflects its capabilities. AutoCAD is the program for professional users – mainly for mechanical and aeronautical engineering and architecture – who will be using large plotters (A3 upwards) and whose investment in hardware and software will be from £10,000 upwards.

One of the considerable advantages of AutoSketch is that its files are compatible with those of AutoCAD, assuming that the version of

AutoCAD is reasonably up-to-date. This allows transfer of work from one to another. The transfer is done by a type of file referred to as DXF (Drawing Exchange Format). These DXF files can also be used to transfer drawings to some DTP programs.

Transferring AutoSketch files to AutoCAD

The AutoSketch file should exist in its standard form as a disk file whose extension letters will be SKD: for example, MTNBIKE2.SKD, ENGINE.SKD.

1. Use *Open* from the *File* menu to read the SKD file from your disk, so that the drawing shows on the screen.

2. Select *Make DXF*, also from the *File* menu. Use the same main filename, or change the name if you want to. You can use a filename that includes another drive or directory.

3. Click on OK to create the file. The message `DXF Out` appears while the file is being created.

4. Use the DXF file in AutoCAD, following the instructions in the AutoCAD manual.

Transferring AutoCAD files to AutoSketch

1. Create a DXF file from your AutoCAD file, following the instructions in the AutoCAD manual.

2. Start AutoSketch, and select *Read DXF* from the *File* menu.

3. Fill in names for *Directory* (such as B:\ or C:\DRAW) and click on OK.

4. Select the filename when it appears (use PageUp or PageDown on a long list).

5. Click on OK when the file is selected.

6. The drawing will appear after some time – the message `DXF In` is at the foot of the screen during the waiting period.

☐ If the drawing is not visible when the `DXF In` message disappears, use *Zoom Full* from the *View* menu to see the drawing. You will probably need to make a new plot box to plot this drawing.

☐ Because AutoCAD contains a large number of features that are absent from AutoSketch, some drawings may not transfer entirely in their original form. The type of drawings that can be made using AutoSketch, if they have been created in AutoCAD, will transfer without problems, but drawings which include AutoCAD features that are absent in AutoSketch, such as 3-D views, will transfer ignoring these features.

Converting drawing files for DTP use

If you need to place circuit diagrams created by AutoSketch into page masters that are being generated by a word processor or a desktop publishing program there are usually several options available.

- The WordPerfect word processor will convert and read SKD files, converting them into the WPG files that can be used as images in WordPerfect.

- Another option is the use of SLD files. AutoSketch contains in its *File* menu the option *Make Slide*. Select this while a drawing is on the screen, and specify a drive/directory in the usual way for the file, which will have the extension letters of SLD. This is a different type of file, one which resembles the files created by Paint-type programs – it specifies position and colours of dots rather than the start and end of lines.

- Your DTP package may be able to read SLD files directly – consult the manual for the DTP package on this – or it may include a converter program which will convert SLD files into one of the more common formats, such as TIF or PCX.

- You can buy converter programs, such as OPTIKS (from the Public Domain Software Library) which can convert SLD files into other formats.

Another way of passing AutoSketch drawings to DTP is to make use of a plotter file. This method works well with Aldus PageMaker, one of the most popular of the 'heavyweight' DTP programs. The method requires the following steps:

1. Reconfigure AutoSketch, selecting *ADI Plotter* (either) and *Binary File* as the way of printing files.

2. When the drawing is on screen, make a suitable plot box.

3. Plot, providing a suitable filename (drive/directory as required).

    This will produce a file with the PLT extension, which can be read by PageMaker.

4. A commercial program – Metafile, from ZenoGraphics – will convert from DXF files into the CGM format, if your DTP package can use CGM files.

Macro facility

AutoSketch contains a macro facility, which amounts to making a recording of a set of actions that can be stored as a file and replayed at any time. This is intended to allow you to automate actions.

☐ The macro facility is useful only when you always work using the same settings. If a macro made using one set of settings is played back when you are using a different set the result will be chaotic. Since the use of macros is not really useful for working with circuit diagrams a description has not been included here.

Appendices

A: Aciran for Windows

The Windows version of Aciran greatly simplifies the work of installation by using an automated INSTALL file. The following description assumes that you have obtained Aciran for Windows on a disk size that is suitable for your A: floppy drive. It also assumes that you are familiar with using Windows for other programs.

1. Start up Windows (which must be Version 3.0 at least, preferably Version 3.1 or higher).

2. Insert the Aciran distribution disk into the A: drive.

3. Type INSTALL and press the ENTER key. Wait until you are informed that installation is complete (or until all disk activity ceases).

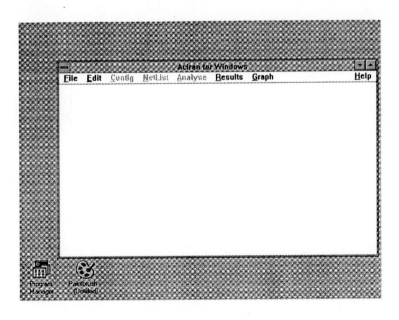

Figure A-1 *The appearance of the opening screen of Aciran for Windows. The similarity with the DOS version is so close that the instructions in Chapters 1 to 6 apply unchanged to this version.*

When you start up Windows there will now be a new Aciran for Windows group icon. Double-click on this icon to see the Aciran group, into which you can place other file sets, such as drawing files.

☐ Initially, all the Aciran files are contained in one directory and in one Windows group.

The appearance of the Aciran for Windows screen is as shown in Figure A-1 above. The menus on the top line follow the same pattern as for the MS-DOS version, so that the description used in the first part of this book applies as closely to this version as to the MS-DOS version.

B: AutoSketch for Windows

AutoSketch for Windows is installed into the machine with Windows working, using the set of distribution disks as follows:

1. Place the first disk into floppy drive A:. Start up the Program Manager of Windows.

2. From the Program Manager *File* menu, click on *Run*. In the box that appears, fill in the filename of :

    A:SETUP

3. Click on OK. The SETUP program will then start to install AutoSketch for Windows.

4. You will be asked what options you want to install. Unless you have firm views (or are short of disk space) opt for all.

5. You will be asked what drive and directory you want to use. The default drive is the hard disk C:, and the default directory is WSKETCH.

    A useful alteration is to use the C:\WINDOWS\WSKETCH path for the directory.

6. You will be asked to fill in your name, company name and the registration number that is printed on the stickers in the package.

7. From then on, installation is automatic and you will be asked to change disks at intervals until installation is complete.

The WSKETCH directory that is created has a set of subdirectories called SUPPORT, TOOLBOX and PARTS, and the PARTS subdirectory has a further subdirectory called CLIP_ART. This latter directory contains a set of ready-made drawings such as maps. Other sets of symbols, such as the symbols you create or buy for circuit diagrams, can be held in other directories branching from the Parts directory.

☐ Note that if you buy sets of symbols these will be U.S. symbols rather than B.S. types.

To start AutoSketch, double-click on the AutoSketch group and then on the AutoSketch icon. When the program starts running, its main screen appears, as illustrated in Figure A-2.

In this screen, the menus of AutoSketch are available as a set of icons, with the *Draw* set the default. Clicking on a menu name at the top of the screen produces the set of icons for that menu, and when the pointer is taken to an icon, the action name will appear at the pointer and also on a line at the top of the screen. There are a few differences in the menus.

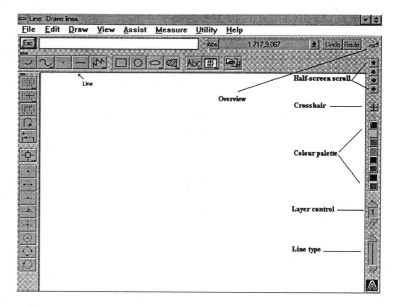

Figure A-2 *The appearance of the opening screen of AutoSketch for Windows, in which menu items are selected by groups of icons.*

1 The *File* menu item called *Print Settings* replaces the *Print Area* item in the MS-DOS version, but the box that it produces is similar.

2 The *Edit* menu now contains *Multi copy*. When an object is selected for copying, it can now be copied to more than one place without the need to select again.

3 There is now a *Utility* menu consisting of *Settings*, *Preferences*, *Toolbox* items, *Macro* items, and file-transfer actions.

4 Some items, such as layer, line type, half screen scroll and colours can be obtained more easily from icons.

5 There is a provision for using the conventional pointer or a large set of cross-hairs as a cursor. The large cross-hairs are convenient as a way of ensuring that objects are lined up.

6 The aircraft icon can be clicked to see an overview. When you are working with a zoomed view, clicking on the aircraft icon will display a small window that shows the full-page view. This action, Figure A-3, can be very slow if the drawing is an elaborate one.

Some actions have also been improved – for example you can open a file by double-clicking on the name rather than by clicking on the name and then on the OK box, as was required in the MS-DOS version.

The set of icons down the left-hand side of the screen allows for commonly-required actions to be carried out more easily than in the MS-DOS version. Figure A-4 shows the meanings of the icons – these are revealed in words when the cursor is taken near.

AUTOSKETCH FOR WINDOWS 271

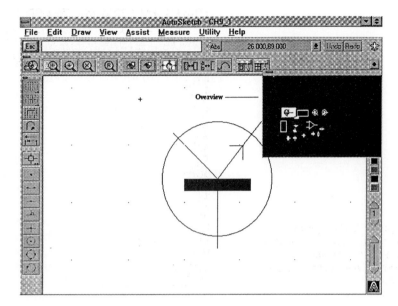

Figure A-3 *The overview action, using the aircraft icon, and used to see a view of the whole drawing while working in a zoomed view.*

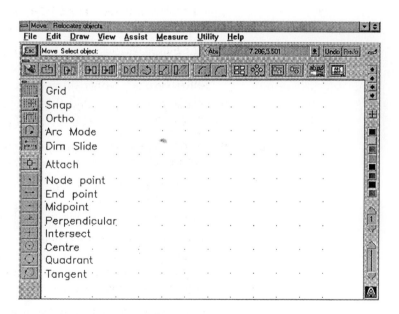

Figure A-4 *The meanings of the icon strip down the left side of the screen.*

☐ Some actions, such as *Move* applied to *Text*, are painfully slow as compared to the MS-DOS version running in the same computer. Others are surprisingly rapid. On the whole, my personal preference is for the DOS version, because I have experienced some oddities in printing old files with the Windows version.

C: The Public Domain and Shareware Library

The PDSL exists to supply disks of programs that are virtually free for inspection, and the only immediate cost to the user is the cost of copying the disks. PDSL can supply on a range of disk formats, and in some cases are virtually the only source of software for some exotic machines. All of the programs are either public domain or shareware. Documentation for each program is included as a disk file, usually with the DOC extension. There are many other suppliers of the same software items, but PDSL was among the first and provides a more full description of the contents of its disks. Like all other reputable PD and shareware dealers, it also warrants, as far as is possible, that none of its software is pirated commercial software. PDSL is a member of the Association of Shareware Professionals.

A public domain program is one for which the author has surrendered all copyright, allowing the program to be copied freely by anyone who wants to use it (the way schools used to copy text books when photocopying came out of an unlimited budget and books came out of a carefully rationed budget).

Many public domain programs are short utilities, and you would normally buy them on a disk that contained 20 - 50 such items. Other PD programs are distinctly longer, and though some of them do not have the polish of a commercial program they must have represented hundreds of hours of effort.

The writers are often professional programmers working at a hobby topic and glad to share the results of their efforts.

Shareware is a rather different concept. The author of a shareware package is hoping to sell directly to the user, cutting out the huge overheads that are involved in having a program manufactured and distributed commercially.

In the early days of shareware, the programs were full working versions, and the poor response by way of payment was a severe blow to authors, particularly in the UK, where users were always less willing to pay for programs than in the USA where the idea started. It has become more common now for shareware programs to be limited to some extent, perhaps running on only a single video card, or unable to use a printer or to create disk files. The user can run the program to a sufficient extent to see if it is likely to be useful, and will have lost very little if it is not.

Registering with the author can be done directly (it is easy to telephone an author in the USA and quote a credit card number) or by way of the PDSL if this can be arranged. The current catalogue contains many programs of particular interest to DTP users, including clip-art, graphics conversion and editing programs, printer utilities, vector-line drawing programs, etc.

Registration can often be done at various levels, with the minimum level entitling you to a copy of the program with all limitations removed. The documentation will be, as for PD items, as a DOC or READ.ME file on the disk. At a higher fee, a full manual is provided and the user is entitled to upgrades at nominal cost. The address for PDSL is:

> Winscombe House,
> Beacon Road,
> Crowborough,
> E. Sussex, TN6 1UL
>
> Tel: (0892) 663298
> Fax: (0892) 667473

At the time of writing, membership subscriptions were £23 per annum for private membership, £73 per annum for corporate membership, and disks were copied for prices of £3.85 each (members) or £4.65 (non-members) with discounts for quantities. There is a surcharge, currently 50p, for 3½" 720 Kb disks. Prices include VAT at the current rate.

Note that some advertisements for shareware list very early versions of Aciran, some as early as Version 1.4. You should avoid sources that deal in such early versions.

Index

3dB point, 42, 48

Abbreviations, 38
Aciran, 12
Aciran disk, 14
Aciran, Windows version, 267
ACNET, 12
Active circuits, 97
Active component, 11
Active delay line, 162
Active device, 29
Active filter, 134
Active low-pass example, 163
ACTRAN, 177
Adding text, 245
Adding transistors, 120
Additional stage, S&K, 144
Algebraic method, 9
Align, 262
Altering LC values, 75
Amplitude, 9
Analyse option, 40
Analysis, principles, 9
Angle, 262
Arc, 202
Arc mode, 257
Area, 262
Array, 227
Arrows, 258
Asymmetrical attenuator, 62
AT computer, 13
Attach option, 195
Attachment modes, 258
Attenuator pad, 58
Audio circuits, 112

Audio circuits, OPamp, 150
AutoCAD, 262
AutoSketch, 185
AutoSketch, Windows version, 269

Backing up, 20
Backslash sign, 18
Bandpass/bandstop filter, 86
Bandpass filter, 160
Bandwidth, 11
Base, part, 208
Baxandall tone control, 122
Bias components, 98
Box, 201
Box, zoom, 212
Box array, 227, 258
Brackets, names inside, 17
Break line action, 204

Cascode, 99
Cassette preamp, 115
CD command, 19
Centre-tapped transformer, 94
CGM file, 264
Chamfer, 258
Change menu, 215
Change option, 50
Change property modes, 256
Change text properties, 250
Characteristics, transistor, 98
Child, 19
Circle, 201
Circuit description, 36
Circuit nodes, 27

Clicking, 190
Clip part, 208
Co-ordinates display, 194
Co-processor, 13, 53, 175
Collector curremt, 99
Color, 259
Compensated attenuator, 64
Computer specification, 13
Configuration, AutoSketch, 188
Constant-k filter, 76
Converting, AutoCAD/
 AutoSketch, 263
Copy, 215
Copying files, 20
Create directory, 18
Crosses box, 208
Ctrl-C keys, 246
Cursor, 34
Cursor choice, 270
Curve, 202, 259
Cut action, Editor, 250

Damping resistors, 83
Data, active devices, 97
Data files, IC, 30
Default printing, 241
Deleted component, 68
Destination disk, 20
Differentiator, Opamp, 135
Dimensions, array, 230
Diode, substitution, 117
Direction of breaking, 205
Directories, 16
Disk size and type, 15
Distance, 262
Dot, 19
Dot filling, 225
Dotted line, 230, 232
Double-dot, 19
Dragging mouse, 191
Drawing example, 233
Drawing limits, 191
Drawing methods summary, 197
Drawing screen, 194
Driving impedance effect, 90
DTP, converting files, 264
DXF file, 186, 262

Editing circuit, 50
Editor, use, 249
Effect of input/output
 impedances, 13
Effect of tolerances, 13
Elaborate ladder network, 70
Ellipse, 202
Ellipse, 259
Elliptical filter, 159
Epson printer, 12
Expanded memory, 185
Export text, 252
Extended memory, 186

Faster plot selection, 242
FETs, 122
File compression, 22
File menu, Aciran, 35
File of output results, 42
Filename, 44
Fillet, 260
Filter in feedback path, 105
Floating transformer, 92
Follow path, 19
Follower, OPamp, 132
Font, 245
Formulae, 10
Frame, 257
Frequency insensitive circuits, 58
Frequency range, 40
Full filename, 46

Gain, 11
Graph printing, 27
Graphical output, 42
Graphics cards, 26
Greyed out item, 191
Grid, 194, 260
Group, 206
Grouping files, 16

Hard disk, 14
Hard disk, Aciran, 25
Height, text, 246
Hercules card, 26, 158
Hewlett-Packard printer, 12
High-pass S&K, 142

Hybrid-pi model, 173

Icons, 239
Impedance, attenuator, 59
Impedance analysis, 107
Impedance converter, 133
Impedance matching, transformer, 94
Impedance of line, 165
Impedances, effect, 13
Import text, 251
Inductor simulation, 147
Inductor symbol, 227
Input, magnetic cartridge, 112
Insert, 250
Insertion of part, 209
Instability, 109
INSTALL program, 22
Installation, AutoSketch, 187
Integrator, OPamp, 134
Inverse display, 190
Inverting OPamp, 129

Ladder network, 67
Laser printer, fine lines, 232
Layers, 254, 260
Layout display, 39
LC filters, 75
Limitations, shareware version, 34
Limits, 233, 260
Limits, AutoSketch, 191
Limits, zoom, 214
Line, 200
Line settings, 232
Line type, 260
Linear plot, 47
List, directory, 18
Load circuit file, 46
Load sketch file, 239
Logarithmic frequency range, 40
Logarithmic plot, 47
Low-pass filter, 9
Low-pass S&K, 140

M-derived filter, 77
Machine font, 250

Macro facility, 265
Matching stub, 165
Measurements, 262
Measuring units, 194
Menu, 34
Menu differences, Windows Autosketch, 270
Metafile, 264
Miller feedback, 117
Mirror image, 217
Model, active device, 98
Models file, 121
Monitor, digital or analogue, 26
Monte Carlo method, 53
Mouse, 34, 190
Move, 215
Moving-coil preamp, 114
MS-DOS filenames, 45
Mu sign, 252
Multi-copy, 270

Negative feedback circuits, 102
New font, 247
Nodes, circuit, 27
Notch filter, 89
Number of tolerance passes, 54
Number One Systems, 12
Numbering nodes, 28

Omega sign, 252
Open circuit line, 167
Opening screen, Aciran, 34
Operational amplifier, 127
ORCAD, 27, 177
Oscillators, 9
Overtype, 250
Overview, 270

Pan action, 214
Paper size, 241
Parallel RC, 11
Parent, 19
Part, 208
Part base, 260
Passive circuit, 9
Paste action, Editor, 250
Pattern, 227, 260

Pattern fill, 221
PC-ECAP, 12
PC compatible, 12
PDSL, 15
Pen-plotter, 240
Phase, 9
Phase-shift graph, 43
Phase-splitting action, 92
Phase angle, 10
Phase diagram, 9
Phase reversal, resonant, 78
Phase reversals, 111
Phase shift in feedback circuits, 106
Phasor diagram, 9
Pi-section, 76
Pick interval, 224, 260
PKUNZIP program, 22
Plot box, 241
Plotter file, PageMaker, 264
Plotting, 240
Pointer, 188
Polyline, 20, 230, 260
Positive feedback, 109
Potentiometer, representing, 123
Power gain, 119
Preparing for drawing, 199
Print area menu, 241
Printer, 12, 27
Printing, 240
Printing graphs, 50
Properties, 256, 261
PSPICE, 179
Public domain, 14, 272

Quicktext, 245

RC filter, 89
README.DOC, 23
Redo action, 203
Reference, component, 36
REGISTER.DOC, 23
Registration, Aciran, 175
Replace option, 50
Resolution, 186
Resonant frequency, 76

Resonant line, 165, 168
Ring array, 230, 261
Root directory, 17
Rotated printing, 241
Rotation, 216
Rubber-banding, 200

Sallen & Key circuits, 140
Save drawing, 238
Save results as file, 46
Saving circuit, 44
Scale, 218
Scale drawing, 186
Scale factor, 232
Scroll bar, 191
Select object, 207
Series-derived shunt fed, 102
Settings, text, 246
SETUP program, 22
Shapes, 199
Shareware, 12, 14, 272
Short circuit line, 167
Shunt-derived series-fed, 103
Simple high-pass example, 33
Simulation, semiconductor, 173
Single-drive machine, 22
SKD file, 239
SKETCH.CFG file, 189
SLD file, 264
Slew-rate, 127
Small circle, 224
Snap, 195, 261
Snap points, 200
Source disk, 20
Special effects, text, 247
Specialised ICs, 128
Specifications, OPamp, 127
Steep-cut filter, 81
Stray capacitance, 99
Stray capacitance, attenuator, 63
Strays, effect on OPamp, 130
Stretch, 219
Summary, Aciran methods, 44
Summary, drawing methods, 197
Summary, editing keys, 252
Symbol set, 211

T-section, 75
Table output, 41
TAC, 175
Template, 199
Terminating resistance, theoretical, 77
Terminating value, filter, 67
Text, 245, 261
Text editor, 251
Text editor, AutoSketch, 245
Thick line, 230
Tick mark, 190
Time delay graph, 43
Tolerances, 13, 31, 52
Tolerances, attenuator, 65
Transformer, 92
Transformer, real, 170
Transistor, 98
Transistors, adding, 120
Transmission lines, 9
Tree, directory, 16
Triplet, OPamp filter, 145
Twin-T circuit, 89
Twin-T, OPamp, 137
Typing text, 247
Typing values, 37, 38

Undo action, 203
Units, 194, 261
Unnamed file, 238
Using co-ordinates, 201
Using dots, 237
Using Editor, 249
Using layers, 254
Utility menu, 270

Values, component, 37, 38
VGA card, 26, 185
Video screen, 185
Voltage gain, 10
VSWR, 166

Wein bridge, 91
Wein bridge, OPamp, 138
Wein bridge with phase splitting, 93
Wide-band amplifier, 117
Winding resistance, 79
Window box, 208
Windows, Aciran, 267
Windows version, 15
Write-protection, 20

XT computer, 13

ZIP files, 22
Zooms, 212